GROWTH CLUSTERS IN EUROPEAN METROPOLITAN CITIES

The European Institute for Comparative Urban Research, EURICUR, was founded in 1988 and has its seat with Erasmus University Rotterdam. EURICUR is the heart and pulse of an extensive network of European cities and universities. EURICUR's principal objective is to stimulate fundamental international comparative research into matters that are of interest to cities. To that end, EURICUR co-ordinates, initiates and carries out studies of subjects of strategic value for urban management today and in the future. Through its network EURICUR has privileged access to crucial information regarding urban development in Europe and North America and to key persons at all levels, working in different public and private organisations active in metropolitan areas. EURICUR closely cooperates with the Eurocities Association, representing about 80 large European cities.

As a scientific institution, one of EURICUR's core activities is to respond to the increasing need for information that broadens and deepens the insight into the complex process of urban development, among others by disseminating the results of its investigations by international book publications. These publications are especially valuable for city governments, supra-national, national and regional authorities, chambers of commerce, real estate developers and investors, academics and students, and others with an interest in urban affairs.

Euricur website: http://www.euricur.nl

This book is one of a series to be published by Ashgate under the auspices of EURICUR, European Institute for Comparative Urban Research, Erasmus University, Rotterdam. Titles in the series are:

The City: Engine Behind Economic Recovery
Edited by Leo H. Klaassen, Leo van den Berg and Jan van der Meer

Governing Metropolitan Regions
Leo van den Berg, H. Arjen van Klink and Jan van der Meer

Urban Tourism
Leo van den Berg, Jan van der Borg and Jan van der Meer

Metropolitan Organising Capacity
Leo van den Berg, Erik Braun and Jan van der Meer

The European High-Speed Train and Urban Development
Leo van den Berg and Peter Pol

National Urban Policies in the European Union
Leo van den Berg, Erik Braun and Jan van der Meer

Growth Clusters in European Metropolitan Cities

A comparative analysis of cluster dynamics in the cities of Amsterdam, Eindhoven, Helsinki, Leipzig, Lyons, Manchester, Munich, Rotterdam and Vienna

LEO VAN DEN BERG
ERIK BRAUN
WILLEM VAN WINDEN

European Institute for Comparative Urban Research
Erasmus University Rotterdam
The Netherlands

LONDON AND NEW YORK

First published 2001 by Ashgate Publishing

Reissued 2018 by Routledge
2 Park Square, Milton Park, Abingdon, Oxon OX14 4RN
711 Third Avenue, New York, NY 10017, USA

Routledge is an imprint of the Taylor & Francis Group, an informa business

A Library of Congress record exists under LC control number: 2001091441

ISBN 13: 978-1-138-73444-9 (hbk)
ISBN 13: 978-1-138-73439-5 (pbk)
ISBN 13: 978-1-315-18729-7 (ebk)

Contents

List of Figures vi
List of Tables viii
Acknowledgements xi

1 Introduction and Background 1

2 Theory and Frame of Analysis 4

3 The Tourist Cluster of Amsterdam 14

4 The Mechatronics Cluster in Eindhoven 39

5 The Telecom Cluster in Helsinki 68

6 The Media Cluster in Leipzig 99

7 The Health Cluster in Lyons 126

8 The Cultural Cluster in Manchester 160

9 The Media Cluster in Munich 189

10 The Audiovisual Cluster in Rotterdam 212

11 The Health Cluster in Vienna 231

12 Synthesis 256

List of Figures

Figure 2.1	Framework of reference	7
Figure 2.2	Inter-organisational relations	11
Figure 3.1	The Amsterdam region	16
Figure 3.2	Population development of Amsterdam (1650–1998)	17
Figure 3.3	Relations in the cluster	29
Figure 3.4	Framework of reference	33
Figure 4.1	The greater Eindhoven region	40
Figure 4.2	Employment structure of the Eindhoven region, 1994–96, number of people employed	42
Figure 4.3	Mechatronics as combination of technical disciplines	44
Figure 4.4	Mechatronics organisation matrix	45
Figure 4.5	Firm typology	48
Figure 4.6	Relations and linkages in the mechatronics cluster in Eindhoven	53
Figure 4.7	Framework of reference	62
Figure 5.1	Major alliances in telecommunications	73
Figure 5.2	Liberalisation of Finnish telecom market, 1882–1998	75
Figure 5.3	Relations in the cluster	85
Figure 5.4	Framework of reference	92
Figure 6.1	Germany, Saxony, Leipzig	101
Figure 6.2	Demand and supply changes in communication	104
Figure 6.3	The media field in Leipzig	108
Figure 6.4	Interaction in the cluster	116
Figure 6.5	Framework of reference	121
Figure 7.1	Health expenditure as % of GDP in different countries, 1980 and 1996	130
Figure 7.2	Locations of medical research activities	141
Figure 7.3	Interactions in the cluster	144
Figure 7.4	Framework of reference	151
Figure 8.1	Manchester in the United Kingdom	161
Figure 8.2	The economic impact of culture in a city or region	166
Figure 8.3	Example of a supply chain in the cultural industries	168
Figure 8.4	The example of design	169
Figure 8.5	Size distribution for cultural enterprise	171
Figure 8.6	Locations within the design cluster	178

Figure 8.7 Relations and dynamics in the cluster 179
Figure 8.8 Confrontation with the research framework 183
Figure 9.1 The spectrum of new audiovisual activities 205
Figure 9.2 Framework of reference 206
Figure 10.1 Sectoral change in employment, 1977 and 1994 214
Figure 10.2 The 100 largest firms in Rotterdam: sector division 215
Figure 10.3 Interaction in the audiovisual cluster in Rotterdam 222
Figure 10.4 Framework of reference 226
Figure 11.1 Location of actors in the health cluster in Vienna 241
Figure 11.2 Relations in the cluster 243
Figure 11.3 Framework of reference 251
Figure 12.1 The 'virtuous circle' of cluster development 263
Figure 12.2 Levels of strategic interaction 265
Figure 12.3 Embeddedness of education 266
Figure 12.4 Integration of research institutes 267

List of Tables

Table 1.1	Some data on the participating cities	2
Table 3.1	Employment in the Amsterdam Region (1995)	18
Table 3.2	Economic development of the Amsterdam region and the Netherlands; relative annual changes	18
Table 3.3	Top 10 of most visited cities in 1991 and 1996	19
Table 3.4	International tourist receipts (in US$ million)	21
Table 3.5	Registered overnight stays in Amsterdam	22
Table 3.6	Cultural attractions in the city centre and Amsterdam	23
Table 3.7	Hotels in Amsterdam (1997)	23
Table 4.1	Distances from Eindhoven to nearby international cities (in kilometres)	40
Table 4.2	Some economic indicators of the Eindhoven region and the Netherlands, 1996	41
Table 4.3	Knowledge-intensive industrial clusters, Eindhoven/Venlo, people employed in different categories, 1996	43
Table 4.4	Largest firms (in terms of employment) in the Eindhoven region	43
Table 4.5	Large firms in the cluster	48
Table 4.6	Philips research centres worldwide	56
Table 4.7	Contributors to the Eindhoven twinning centre	60
Table 5.1	Population in Helsinki and Helsinki region	69
Table 5.2	Jobs by industry in Helsinki, metropolitan area, region and Finland, 1996, in % and absolute totals	71
Table 5.3	Finland leading in mobile phones and internet penetration	75
Table 5.4	The telecommunications sector in the Helsinki metropolitan area, 1996	77
Table 5.5	University level education and number of students in the Helsinki metropolitan area, 1997	80
Table 6.1	Sector structure of the Leipzig economy, 1997	102
Table 6.2	Segmentation of on-line services	106
Table 6.3	Number of firms, employment and turnover in Leipzig's media cluster, 1996	108
Table 6.4	Education institutes in media, Leipzig	113
Table 6.5	Contribution of participants in media stimulation	115
Table 7.1	Population in 1996 and development 1982–90	127

Table 7.2 Evolution of total number of employees in 10 urban regions in France, between 1989 and 1996 128
Table 7.3 Structure of the economy of Aire Urbaine de Lyon, 1996 128
Table 7.4 Employment in industrial branches, Air Urbaine de Lyon, 1997 128
Table 7.5 Public part of health expenditures in %, 1994 131
Table 7.6 The biotech industry in USA and Europe 1997, some indicators 132
Table 7.7 Large pharmaceutical firms in Lyons 134
Table 7.8 Size distribution of producers of medical devices, Rhône-Alpes region 135
Table 7.9 Students in Lyons 140
Table 7.10 Medical students 142
Table 7.11 Number and value of contract activities intermediated by EZUS, 1990–97 146
Table 7.12 Distribution of contract research in 1995–96 146
Table 7.13 What the actors in the cluster can offer each other 156
Table 8.1 Employment in the Manchester TEC area for 1995 164
Table 8.2 Economic performance of the Manchester TEC area in perspective 164
Table 8.3 Manchester cultural enterprise in UK perspective 166
Table 8.4 Examples of cultural employment growth in some countries 167
Table 8.5 Employment and firm distribution in the subsectors of the cultural sector 171
Table 8.6 Students at Manchester institutions 174
Table 9.1 Employees registered in social security system by sector in the Munich region, city and surrounding area, 1997 191
Table 9.2 Number of students in Munich, 1997 191
Table 9.3 Some statistics on the media sector in Munich 192
Table 9.4 Employment audiovisual activities (TV and radio) in the main German media cities, 1995 195
Table 9.5 Private broadcasting: turnover in millions of DM and shares of major cities, 1993 196
Table 9.6 Large firms in media and communication, 1995 200
Table 9.7 Pilot projects of Bavaria Online 202
Table 9.8 Some public-private media-related projects in the Munich region 209
Table 10.1 Economic structure of Rotterdam, 1996 213

Table 10.2	Firms in Rotterdam, sectors and size classes, 1996	214
Table 11.1	Industry in Vienna: employees and added value in 1993, % change 1981–93	233
Table 11.2	Economic structure of Vienna, 1973, 1986 and 1994	233
Table 11.3	Employees in the health cluster	234
Table 11.4	Large pharmaceutical firms in Vienna	235
Table 11.5	University of Vienna: number of students and graduates in several fields, 1996/97	238
Table 11.6	Figures on public hospitals and nursing homes of the Wiener Krankenanstaltenverbund, 1997	240

Acknowledgements

In 1997, the European Institute for Comparative Urban Research (EURICUR) was invited by members of the Economic Development and Urban Regeneration Committee (EDURC) of Eurocities to carry out an international comparative study of growth clusters and urban economic development. During the investigation period, several meetings were organised with the participating cities, to feed the findings back and to exchange experiences. This book contains the findings of the investigation. It sheds new light on growth processes and mechanisms in European cities, and offers much scope for local and regional governments to pursue effective policies. The main part of the information needed for the cluster analysis was generated by interviewing representatives of municipalities, firms, educational institutes and other organisations related to the cluster under consideration. Without their help and information, the investigation could not have been accomplished at all. Therefore, we want to express our great appreciation of the welcome we received in each of the cities involved in the research.

The research has been carried out under supervision and guidance of the city of Munich, represented by Dr Klaus Schussman. We thank him warmly for his commitment, his enthusiasm, his valuable contributions and, last but not least, his 'organising capacity', which proved indispensable to carry out a project with so many cities and people involved. Next, we would like to thank the representatives of the participating cities. They were of great help in the organisation of the visits and the feed-back of the results to the discussion partners, and contributed with valuable remarks. We thank the following people for their pleasant cooperation (in alphabetical order): Mrs Lyn Barbour (Manchester), Mr Helmut Bernt (Leipzig), Mrs Gabrielle Bock (Leipzig), Mr Huub Bouman (Eindhoven), Mr Dave Carter (Manchester), Mr Bernhard Eller (Munich), Mr Jos Geerling (Rotterdam), Mr Eero Holstila (Helsinki), Mrs Claudia Hörter (Munich), Mr Peter Nagel (Eindhoven), Mr Thomas Resch (Vienna), Mr Ilkka Susiluoto (Helsinki), Mr Pierre-Yves Tesse (Lyons), Mr Reinhart Troper (Vienna) and Mr Rob Wijnveen (Amsterdam).

We thank our colleague Dr Arjen van Klink for his contribution to the case of Lyons, as well as for his valuable ideas and suggestions for the analysis of clusters in general. We thank another colleague, Dr Alexander Otgaar for his contribution to the case study of Amsterdam. Naturally, only the three authors are responsible for the contents of the book. Finally, we would like to thank Mrs Attie Elderson-de Boer for her corrections, and the Euricur secretaries Arianne van Bijnen and Ankimon Naerebout for their indispensable help during the investigation.

Leo van den Berg
Erik Braun
Willem van Winden

Chapter One

Introduction and Background

1 Introduction

Sustainable economic growth is of high interest to European cities: it is indispensable to further the well-being and prosperity of citizens and firms, and to generate employment. Thus, it is important to gain insight into the economic growth opportunities in cities. In this respect, new growth sectors such as information technology, biotechnology, environmental technology and tourism are at the centre of interest to academics as well as to urban managers. Many cities invest heavily in developing and attracting industries in these promising sectors. However, little is known about critical success factors that determine economic development of cities and regions and empirical studies that draw lessons for policy are scarce (Nijkamp, 1999). Moreover, there are good reasons to doubt to what extent a purely sectoral view is adequate to analyse urban economic growth and to design policies. There are many indications that urban economic growth increasingly seems to emerge from fruitful cooperation between economic actors, who form innovative networks. It is in these geographically concentrated network configurations, or 'clusters', that value-added and employment growth in urban regions is realised. This demands a new policy approach in urban economic development. The general aim of this study is to increase the insight into new growth opportunities for European cities and to provide scope for urban policy.

2 Background

This book is based on the results of international comparative research in nine European cities into growth clusters and the scope for urban economic policy (Berg, Braun and Winden, 1999). The investigation has been conducted among member cities of the Eurocities-network. The following cities were surveyed (in alphabetical order): Amsterdam (The Netherlands), Eindhoven (The Netherlands), Helsinki (Finland), Leipzig (Germany), Lyons (France), Manchester (UK), Munich (Germany), Rotterdam (The Netherlands) and Vienna (Austria). The cities differ in size as well as in economic structure and performance, as can be seen in Table 1.1

Table 1.1 Some data on the participating cities

City	Inhabitants of agglomeration	GDP/capita (in ECU, 1995)**	Chosen cluster
Amsterdam	1,300,000	12,505	Tourism
Eindhoven	670,000	n.k.	Mechatronics
Helsinki	920,000	16,441	Telecom
Leipzig	502,878 *	n.k.	Media
Lyons	1,262,000	13,189	Health
Manchester	2,591,000	11,079	Cultural industries
Munich	1,241,000	17,268	Media
Rotterdam	1,065,000	13,341	Media
Vienna	1,807,000	18,649	Health

Sources: * Leipzig City Council, 1996; ** Ereco, 1997 in Mayerhofer and Palme, 1996.

In each of the cities one potential growth cluster has been selected on the basis of growth figures or the city's growth expectations and opportunities. The health cluster – the complex of health care institutes, medical and biological research, the pharmaceutical industry, medical instruments – was investigated in the cities of Lyons and Vienna, both with a great tradition in medical research and health care. Both cities share an ambition to make more out of their medical complexes in economic terms. In Munich, Rotterdam and Leipzig we have studied the complex of media and related activities as growth clusters, although there were important differences. In Munich, the media cluster is very large and very well developed. In Rotterdam, the media industry is very small, but the municipality considers this cluster an important element in their strategy to diversify the city's economic base and to create new employment. For Leipzig, the situation is again very different: as a former GDR city, Leipzig seeks to re-establish the media cluster in which it had a great tradition. For Helsinki, the investigation was concerned with the cluster of telecommunications – both the production of equipment and services – as this cluster is characterised by very high growth rates, with Nokia, a world leader in mobile phones, playing a very important role. In Eindhoven the mechatronics cluster, a high-tech industrial cluster, was surveyed. For Amsterdam, tourism was the target cluster. In Manchester, finally, we investigated the cultural industries as growth clusters, with a special eye for the potential for urban regeneration.

3 Organisation of the Book

This book is organised as follows. Chapter Two provides a theoretical background, as well as a framework of reference to analyse the clusters in the nine cities involved. Chapters Three to Eleven contain the case studies of the cities involved: Amsterdam (Chapter Three), Eindhoven (Chapter Four), Helsinki (Chapter Five), Leipzig (Chapter Six), Lyons (Chapter Seven), Manchester (Chapter Eight), Munich (Chapter Nine), Rotterdam (Chapter Ten), and Vienna (Chapter Eleven). The case studies are based on extensive literature study, internet resources, and in-depth interviewing: in each of the cities, many interviews with key persons were held to obtain the right information. Chapter Twelve finally contains the synthesis of the findings: the analysis of the cases is fed back to the framework of reference, resulting in general conclusions and recommendations about growth cluster development in European urban regions.

References

Berg, L. van den, E. Braun and W. van Winden (1999), *Growth Clusters in European Metropolitan Cities: A new policy perspective*, European Institute for Comparative Urban Research, Erasmus University, Rotterdam.

Leipzig City Council (1996), *Leipzig: Facts and figures.*

Mayerhofer, P. and G. Palme (1996), *Wirtschaftsstandort Wien: Positioniering im Europäischen Städtenetz*, WIFO, Wien.

Nijkamp, P. (1999), 'De Revival van de Regio', in W. de Graaff and F. Boekema (eds), *De Regio Centraal*, Van Gorkum & Comp., Assen.

Chapter Two

Theory and Frame of Analysis

1 Introduction

This chapter sketches a background for the analysis of growth clusters of European cities and develops the framework of reference that is to be used in the case studies. First, in section 2 the cluster perspective is introduced, as a useful tool to analyse new growth in metropolitan regions at a time when networks play such a dominant role as organising principle. Second, on the basis of the issues raised, a framework of reference is presented in section 3. This framework serves as a basis for the analysis of urban economic growth sectors in the participating cities, and as a tool to design policies aimed at fostering sustainable growth. Section 4 elaborates on the methodology that was used to put the framework into practise in the case study analyses.

2 Urban Growth, Networks and Clusters

Since the early 1990s networks have been recognised as a very important ordering principle in the Western world's economies. Firms and organisations more and more actively engage in networks as a means to survive in a volatile international market and to cope with rapid technological change. Engagement in networks has several well-documented advantages (Jarillo, 1993; Castells, 1996; and many more). It makes for flexibility: to benefit from chances, a firm has to be able to react fast, and to engage in partnerships with complementary strengths and capabilities. Networks are particularly important regarding innovation. Strong international competition and fast technological development urge firms to innovate constantly in terms of producing new products or services, developing new processes and accessing new markets. Participation in a network enables a firm to concentrate on core capabilities, and provides access to resources (such as specific know-how, technology, financial means, products, assets, markets etc.) in other firms and organisations. This helps them to improve their competitive position.

Interfirm and inter-organisational cooperations in networks have different spatial dimensions. Networks can extend worldwide, as do the global networks of stock exchanges and financial markets. But many network relations between actors can be located in a specific area, region or city. The popular term 'cluster' is mostly related to this local or regional dimension of networks. In the literature, clusters are defined and described in many different ways (see Porter, 1990; Berg, Klink and de Langen, 1997; Jacobs, 1996; Lazonick, 1992 and many others), but most definitions share the notion of clusters as localised networks of specialised organisations, whose production processes are closely linked through the exchange of goods, services and/or knowledge. Particularly the informal exchange of information, knowledge and creative ideas is considered an important characteristic of such networks. Unlike a sector, a cluster unites companies from different levels in the industrial chain (suppliers, customers), with service units (financial institutions, production-supporting services) and with government bodies, semi-public agencies, universities, research institutes etc. Many authors have stressed the dynamics of clusters. Already in 1927, Marshall described the powerful dynamics in industrial districts, where geographically concentrated groupings of firms, large and small, interact with each other via subcontracting, joint ventures or other collaborative means, gaining external economies of scale in doing so (Cooke, 1995), thus deriving international competitiveness from local sources. Porter (1990) describes how clusters of densely networked firms serve global markets while deriving their strength from a regional basis. He discerns four conditions as essential in that development: factor conditions (quality of labour, capital, knowledge available); demand conditions (scale and quality of the regional home market); supplier industries (globally competitive suppliers, specialised services); and business strategy (rivalry between local firms but also willingness to cooperate in research, sales and marketing). In particular, the interplay of competition and cooperation is fundamental. Too much competition may be destructive, but the same holds for too much cooperation when it degenerates into the formation of cartels (Cooke, 1995; Harrison, 1994). Lazonick (1992) and Boekholt (1994) stress that in clusters, a major role is played by other than interfirm linkages: links with government-supported scientific institutes, ties with the scientific community and professional associations are important factors in a clusters' performance. Yet the question remains why proximity still seems to matter in networks, where modern communication technology theoretically permits spatial dispersion. Several reasons are put forward. First, face-to-face contacts appear to be very important as sources of (technological) information and in the exchange of tacit knowledge (Leonard-Barton, 1982; Malmberg et al., 1996). Spatial proximity greatly enhances the possibility of such contacts. Second, cooperation between actors requires mutual trust. This holds particularly when sensitive and valuable information is exchanged, for instance in a joint innovation project. Several authors (Piore and Sabel, 1984) argue that cultural proximity, i.e. the sharing of the same norms and values, is an important factor in that respect, since cooperation is a human phenomenon. A very relevant issue concerning the spatial dimension of clusters is how local networks relate to global networks. In the local-global interplay, transnational companies (TNC) play a special role. If a TNC is rooted and integrated ('fledged') in the region and engaging in regional networks, it can act as an important disseminator of new knowledge, information and innovation from abroad into the region. This is particularly relevant for research and development activities: knowledge flows are facilitated by personal relationships and mobility of employees (Malmberg et al., 1996) or spin-outs from the large firm.

3 Frame of Analysis

The literature on clusters is extensive. Most studies focus on theoretical aspects of clustering, or take (very) large regions as geographical unit. In empirical studies, there seems to be a bias towards well-performing regions (the 3rd Italy, Baden Würtemberg, Silicon Valley, Route 128–Boston, Cambridge) with high rates of growth and innovation, and dense network structures. However, empirical (comparative) cluster-studies on urban regions are scarce. In our study, we aimed to study clusters in urban regions in an integral way, from the view that clusters are embedded in the spatial-economic and cultural and administrative/political structures of the urban region. We have drawn up a frame of analysis to take several aspects into account and study their inter-relations. Using this framework, we aim to understand growth processes in clusters in urban regions, provide scope for policy improvement and allow the comparison of different types of clusters. It leans heavily on existing literature (partly discussed in the preceding section) and recent insights in the importance of 'organising capacity' in urban regions (Berg, Braun and Meer, 1997).

The framework contains the following, interrelated elements: general spatial-economic conditions; cluster specific conditions; and organising capacity. The general conditions of the urban region comprise the economic and spatial structure, quality of life, and cultural aspects. The cluster specific conditions comprise the quality, size and number of the actors in the cluster, and the prevalence and quality of strategic interaction within the cluster and external linkages. Finally the organising capacity concerns the presence of a vision and strategy regarding the cluster, the level of public-private cooperation, the levels of political and societal support for cluster development and the presence and quality of leadership. Figure 2.1 shows the components of the framework and the interrelations between the parts. In the following, the contents of the framework are elaborated.

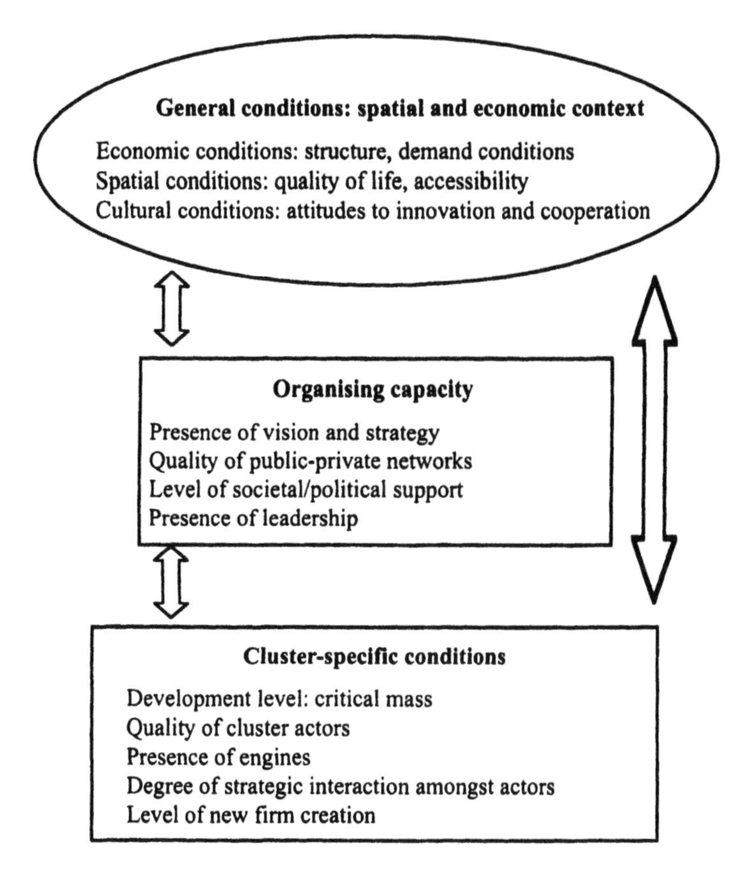

Figure 2.1 Framework of reference

3.1 General Conditions: The Economic, Spatial and Cultural Context

The development of a cluster cannot be understood without a thorough insight into the characteristics of the functional urban region as a whole. In our research, we have distinguished three elements in the spatial economic context: economic, spatial and cultural conditions. The first economic condition is the structure of the local economy. In general, the sector-mix in an urban region influences the development of the specific growth-cluster under consideration, because there are often many interrelations (suppliers-links, demand, services) between the growth sector and other sectors in the urban region. In particular, demand conditions are fundamental to the functioning of a cluster. Who are the buyers of the products/services that the cooperating firms in the cluster produce? Depending on the cluster it may be large companies in the region, that indirectly act as stimulators of the cluster as main client. But governments or individuals are also important clients for some clusters. The demand can be mainly local, but may also come from other regions. Another important characteristic of the urban region is the knowledge base of the local economy, as reflected in the education level of the workforce, the knowledge-intensity of the economic activities that take place in the urban region and the presence of research institutes. The assumption is that in most new growth clusters, knowledge is the key driving force behind growth and development. Therefore, regions with a well-educated workforce and high-quality knowledge institutes will have a general advantage over other regions.

The spatial conditions in an urban area form the second 'context element' of our analysis. Two main aspects of the spatial conditions can be discerned. The first is the quality of life in the urban area. In general, quality of life is a location factor of utmost importance. In contrast to the past, when workers migrated to places where factories and other firms were located, in our modern economy firms seem to move to areas where they can find the appropriately skilled people. Highly skilled people, on whom urban development is strongly dependent, attach much value to a high quality of the living environment, so in an indirect sense the quality of the living environment is an essential factor in economic urban development. A second aspect of the spatial conditions relates to accessibility, in the broadest sense of the word. Accessibility can be split into physical and electronic, internal and external to the urban region. Accessibility is a necessary condition for urban development, since in a network economy, interaction is the key. Bad transport systems in an urban area may seriously hamper interaction in a cluster, particularly if cluster elements are dispersed. External accessibility to other cities and regions is also fundamental, as a means to link local networks up with national and international networks of all kinds. At the same time, the level of external accessibility influences the competition between cities for the development of clusters.

Finally, the 'cultware' is considered an element of the cluster context as institutional variable. Cultware relates to attitudes of people and firms. In particular, attitudes towards innovation are important, for often in growth sectors, the main driving force of the development of the cluster is innovation. Of equal importance is the willingness of people in the urban region to cooperate. Cooperation is also one of the main sources of innovation, new combinations, and hence, growth and development of the cluster.

3.2 Cluster-specific Conditions

Two important aspects of a cluster's functioning are its size and its development level. Does the cluster possess 'critical mass'? How many companies and educational and research institutions are active in the cluster? Critical mass is important for various reasons. First, it ensures a market large enough to support the (specialist) activities in the cluster. A second advantage is that the presence of many companies may invite keen competition and thus force companies to operate efficiently and effectively. Third, the chance of fast penetration of all types of innovation is the greater as the cluster is larger. Fourth, regional cooperativeness is easier to accomplish within a large cluster, as it is easier to find a complementary partner in the region. Finally, scale offers prospects for the sharing of resources, the benefits of a shared pool of specialised labour and the scope for a cluster 'superstructure' like joint education facilities. Next, the quality of the cluster actors is a relevant factor. Quality may refer to the degree of international competitiveness of firms, the technological sophistication of their output, the standing of a university, etc. The presence of one or more engines in a region – be they large multinational firms or other actors – is also supposed to be a determinant of a cluster's functioning, in their role of spider in global and local networks, or as 'flagships' of the cluster as whole. Besides its scale and quality, the degree of strategic interaction is assumed to be largely decisive for a cluster's performance. Strategic interaction implies long-term relations other than strictly financial, between organisations. Within the region such interaction can be achieved on various levels: amongst companies, between companies and institutions of education or research, amongst educational institutions, etc. As indicated in the last section, strategic interaction can serve a variety of purposes: to create scale, to use one another's knowledge (of markets, technology, organisation), to make use of one another's networks, to solve common problems together, or to enhance flexibility. Fundamental conditions for interaction are that actors involved should know (of) one another and trust each other. Next, the parties need to be to some extent complementary. This last aspect is also associated with the scale. In a large and varied cluster, the chance of finding a suitable partner is considerably greater than in a small one. A final element determining cluster dynamics is the level of new firm creation. Young firms are often dynamic and innovative, and generate jobs; they can be important for large firms as partners in innovation, or as suppliers. They may help to tie young talent to the region, particularly when new firms are strongly linked up in the cluster, for instance by strategic relations with local universities or large firms. The creation of new firms in European cities generally lags behind the figures of the USA, particularly of high-tech starters. Appropriate public-private structures to guide starting firms are assumed to be a very important factor in the degree to which people are inclined to start businesses, but also cultural elements (the level of 'entrepreneurial spirit') are likely to play a role.

3.3 Organising Capacity

The final element that presumably plays a part in the performance of the cluster is the degree of organising capacity regarding the cluster. Organising capacity can be defined as the ability of the urban region to enlist all actors involved in the growth cluster, and with their help generate new ideas and develop and implement policy designed to respond to developments and create conditions for sustainable development of the cluster (Berg, Braun and Meer, 1997, adapted). Organising capacity can refer to the development of cluster-specific policy, the attraction of cluster-supporting elements (companies), investment in specific infrastructure, etc. Berg, Braun and Meer (1997) distinguished some elements necessary to the organising capacity in general: leadership, vision/strategy, political/societal support, and public-private partnership. All these elements are important for the development of a cluster in a city or region. Leadership can play a prominent role in developing and stimulating a cluster, especially if many interests are involved and decision-making powers fragmented or nebulous. In many cases, one or a few persons can achieve much. Leadership need not always be vested in the public sector: a large company or other organisation can take the lead in setting up networks in the region and initiating policy. The development of the growth cluster should be steered by an integral vision, and preferably be laid down in a strategy. A well-defined and shared vision and strategy on the development possibilities of a cluster is indispensable for an efficient allocation of resources and efforts to stimulate the cluster. Political and societal support are necessary conditions for a cluster policy as well. Political support helps to bring about positive collaboration on the local level. Proper presentation and communication of policies are of paramount importance to achieve results. Societal support is important for the acceptance of policies aimed at growth clusters. Finally, public-private cooperation – on the strategic, tactical and operational levels – is very important for a successful cluster policy. An essential factor for success is the early involvement of the private sector in the development of locations, the attraction of companies, etc. The knowledge, expertise and involvement of the private sector can be very valuable to the decision-making process, and considerably enhance the chance of success. Besides, government can act as network broker, stimulating the formation of inter- and intra-sectoral networks, by bringing people and firms together. Local or regional government can engage in public-private partnerships directed at the stimulation of the growth cluster, for example by providing facilities or specific education.

4 Methodology

In our view clusters are 'localised' networks of organisations that are closely linked through the exchange of goods, services, people and knowledge. To study clusters empirically, we started with some key actors in the sector that then formed the basis of the analysis of networks in the region: we have studied how these key organisations are linked up with other organisations (firms, knowledge institutes, government) within and outside the region. Basically we have tried to collect some indicative evidence of strategic relations between these actors, as depicted in Figure 2.2.

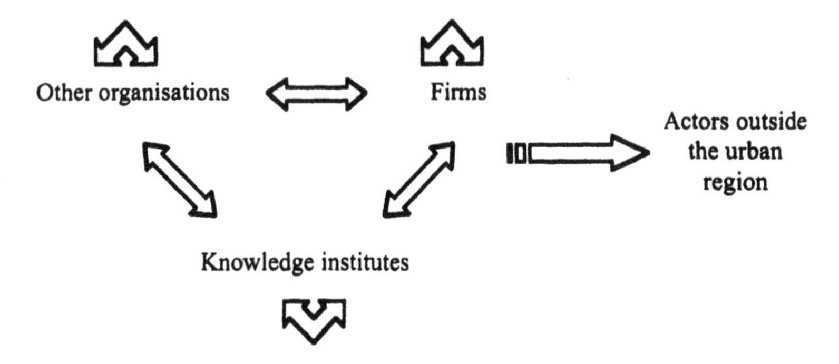

Figure 2.2 Inter-organisational relations

That type of evidence consists of the presence of formal cooperative structures, joint facilities or joint projects in the growth-cluster in the urban region. The research results of the case studies were based on a combination of desk and field research. Extensive use was made of existing papers about the different clusters in the cities. On that basis, interviews were held with key persons in the cluster. These interviews proved to be an indispensable and very rich source of information, particularly for discovering relations between actors in the cluster. In our survey, we analysed several type of clusters: two mature health clusters (Lyons and Vienna); one very small media cluster (Rotterdam); a developing media cluster (Leipzig); a very mature cluster (Munich); a large tourist cluster (Amsterdam); a specialised cultural cluster (Manchester); and two mature technologically-oriented clusters (telecommunications in Helsinki and mechatronics in Eindhoven). At first sight, comparison seems difficult: the cases are dispersed across several countries, entailing country-specific aspects; they differ in type, and in their 'development stage'. Despite these wide differences, in Chapter Twelve efforts are made to match the theoretical reflections with the empirical evidence from the nine case studies, and to draw some general conclusions and lessons about new growth opportunities and implications for urban economic policy.

References

Berg, L. van den, E. Braun and J. van der Meer (1997), *Metropolitan Organising Capacity*, Avebury, Aldershot.

Berg, L. van den, E. Braun and W. van Winden (1999), Growth Clusters in European Metropolitan Cities: A new policy perspective, European Institute for Comparative Urban Research, Erasmus University Rotterdam

Berg, L. van den, H.A. van Klink and P. de Langen (1997), *Maritieme Clustering in Nederland*, Erasmus University, Rotterdam.

Boekholt, P. (1994), *Methodology to Identify Regional Clusters of Firms and Their Needs*, paper for Sprint-RITTS workshop, Luxemburg.

Castells, M. (1996), *The Rise of the Network Society*, Blackwell Publishers, Cambridge.

Cooke, P. (1995), 'Regions, Clusters and Innovation Networks', in P. Cooke (ed.), *The Rise of the Rustbelt*, UCL, London.

EDURC (Eurocities Economic Development and Regeneration Committee) (1997), *Growth Sectors in European Cities*, Munich.

Harrison, B. (1994), *Lean and Mean: The changing landscape of corporate power in the age of flexibility*, New York, Basic Books

Jacobs, D. (1996), *Het Kennisoffensief: Slim concurreren met kennis*, Samson, Alphen aan den Rijn.

Jarillo, J.C. (1993), *Strategic Networks: Creating the borderless organisation*, Butterworth-Heinemann, Oxford.

Lazonick, W. (1992), *Industry Clusters versus Global Webs*, Department of Economics, Columbia University, New York.

Leipzig City Council (1996), *Leipzig: Facts and figures*.

Leonard-Barton, D. (1982), *Swedish Entrepreneurs in Manufacturing and their Sources of Information*, Center for Policy Application, MIT, Boston.

Malmberg, A., O. Sölvell and I. Zander (1996), 'Spatial Clustering, Local Accumulation of Knowledge and Firm Competitiveness', *Geografiska Annaler*, 78b (1992–2).
Mayerhofer, P. and G. Palme (1996), *Wirtschaftsstandort Wien: Positioniering im Europäischen Städtenetz*, WIFO, Wien.
Nijkamp, P. (1999), 'De Revival van de Regio', in W. de Graaff and F. Boekema (eds), *De Regio Centraal*, Van Gorkum & Comp., Assen.
Piore, M.J. and C.F. Sabel (1984), *The Second Industrial Divide, Possibilities for Prosperity*, Basic Books, New York.
Porter, M. (1990), *The Competitive Advantage of Nations*, The Free Press, New York.

Chapter Three

The Tourist Cluster of Amsterdam[1]

1 Introduction

Amsterdam is the fourth most visited tourist destination in Europe, with more than six million registered overnight stays a year. The number of visitors to the Dutch capital has shown an increase over the last few years at an annual growth rate of 6 per cent. Not surprisingly, tourism is very important to the economy of Amsterdam and to employment in particular. The tourist cluster of Amsterdam consists of an impressive range of tourist attractions, mainly concentrated in the historic city centre with its typical canal structure. Moreover, many people earn their wage in shops, restaurants and pubs; services with tourists as important part of their customers.

This case study aims at describing and analysing the different actors in the tourist cluster of Amsterdam, as well as their mutual relations, and the dynamics of the cluster. Furthermore, the cluster is put in the broader context of the economic, social and spatial structure of the Amsterdam region. To that end, section 2 first sketches a profile of this region. After that, section 3 discusses the relevant characteristics and developments of tourism in general. Next, section 4 describes the various actors of the cluster and their spatial spread over the city. Section 5 analyses and judges the strategic relations between those actors. This analysis will be confronted with the framework of Chapter Two in section 6. Finally, section 7 concludes.

2 Profile of the Amsterdam Region

Amsterdam is one of the four major cities in the Netherlands next to Rotterdam, The Hague and Utrecht. These four cities are the main pillars of the so-called Randstad Holland. This is not an administrative entity, but is considered a centre of international economic importance comparable to London, Paris and the Rhein-Ruhr region. With a population of 718,000, Amsterdam is the largest city of the country.

The city of Amsterdam has 13 town boroughs each with their own council. The area known as the 'Oostenlijke havengebied' (harbour area) comes under the responsibility of the city administration. The city centre, the area encircled by the seventeenth century canals, is also directly governed by the City Council of Amsterdam. Although the city centre of Amsterdam has lost some of its importance in the last few decades, it is still the largest concentration of economic activity in the Netherlands.

However, the inner city is not the only important economic centre in the Amsterdam region. In the so-called ROA-area (Regional Body of Amsterdam) with a population of more than 1.3 million, a number of 'new' economic centres can be identified, such as the Schiphol area, the South Axis and Amsterdam Southeast.

2.1 Population and Social Circumstances

Between 1650 and 1970, the population of Amsterdam increased from 100,000 830,000. From the 1970s onwards the number of inhabitants decreased in a period of suburbanisation and disurbanisation to 676,000 in 1985. Recently, population has stabilised at 720,000. About 80,000 people live in the city centre of Amsterdam. A remarkable development is the rising popularity of this part of the town as place to live. Because the demand for housing far exceeds supply, house prices have risen, attracting mainly high-income groups. Moreover, the city-centre population is characterised by a small proportion of non-natives (compared to other parts of Amsterdam) and a large proportion of high-skilled people.

Social problems are not concentrated in the city centre, but in other town boroughs such as Amsterdam Southeast. The housing in this part of Amsterdam (mainly large apartment buildings) was built in the 1960s and 1970s, and seems unable to attract high-income groups, which has resulted in high concentrations of low-skilled, often unemployed people. The city of Amsterdam tries to relate engines of new economic growth to social revitalisation projects in the context of the (national) Major Cities Policy. The Amsterdam ArenA, a multifunctional stadium located in Amsterdam Southeast, is such a growth engine.

2.2 Economy

Amsterdam has been an important trade city since the Middle Ages. Economic success reached its climax in the seventeenth century, when the magnificent network of canals was laid out. In this so-called 'Gouden Eeuw' (Golden Age), culture flourished alongside business, while artists like Vondel and

Rembrandt created their immortal works. After the seventeenth century, however, the position of Amsterdam in international trade weakened somewhat. In the nineteenth century, industrialisation changed the urban economic structure. The industrial and financial sectors gained more and more importance and Amsterdam became a city of diamond merchants. From the 1970s onwards, the urban economic structure radically changed thanks to the rise of the service sector and the massification of tourism. Today, Amsterdam is generally acknowledged as the financial, cultural and tourist centre of the Netherlands. Services dominate the economic structure of Amsterdam region, accounting for 84 per cent of the total employment (total employment is 660,000). The share of the hotel and catering industry, 5 per cent, is relatively high compared to other cities. Other economic activities of importance are trade and financial and commercial services (see Table 3.1).

Many national and international companies have their headquarters in Amsterdam. Economic activities are fairly equally distributed across the region, with concentrations in the city centre (a tourist, culture and financial centre), Southeast (a 'no-nonsense' business district), the South Axis (a top location) and the Schiphol area (a distribution centre), as depicted in Figure 3.1. Schiphol, the international airport of Amsterdam, is located southwest of Amsterdam. With more than 31 million passengers a year, it is Europe's fourth largest passenger airport. In the last few years, the number of passenger movements has grown at a greater rate than at other airports in Europe

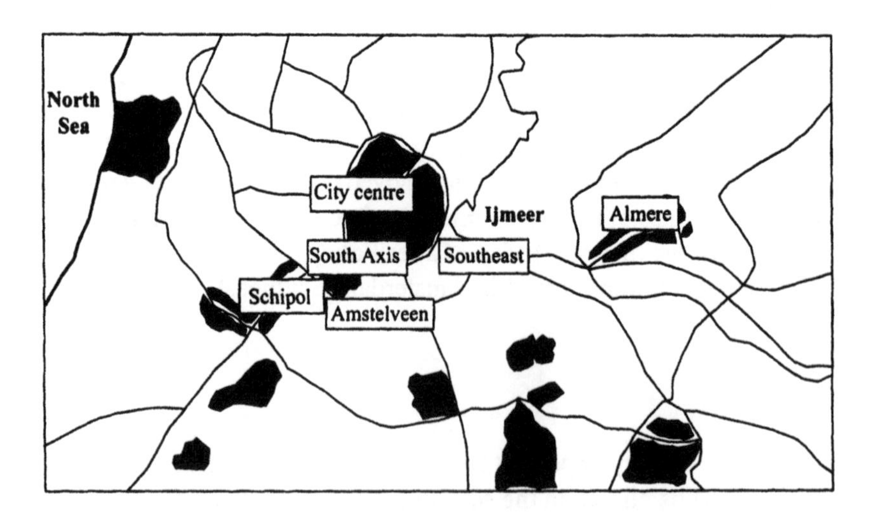

Figure 3.1 The Amsterdam region

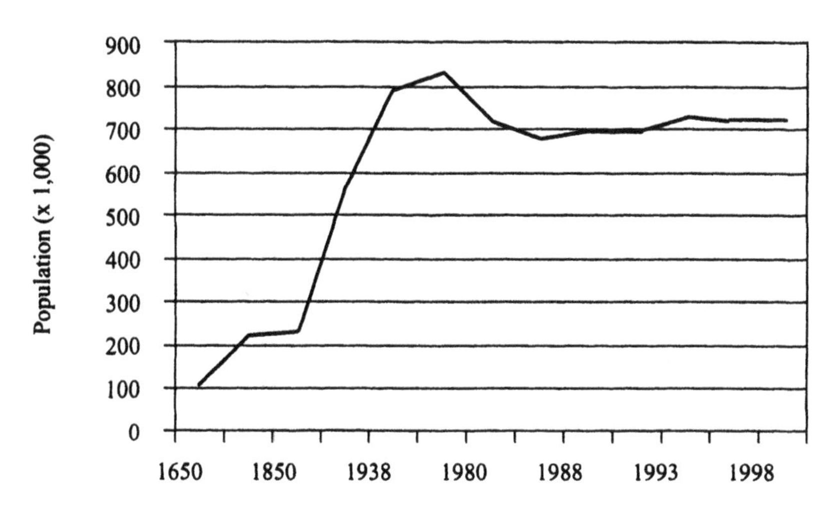

Figure 3.2 Population development of Amsterdam (1650–1998)

Source: Berg and Bramezza, 1992; City of Amsterdam/O+S, 1998.

(Amsterdam Airport Schiphol, 1998). Schiphol provides connections to 220 destinations in 90 countries. Another important pillar of the Amsterdam economy is its port, which handles 55 million tonnes of cargo each year. The passenger cruise industry is an important growth market for the port of Amsterdam. Every year, about 60,000 tourists arrive by one of the hundred cruise vessels that call at the city (AMPRO, 1998).

In the last few years, the economy of the Amsterdam region has grown at a higher rate than the national economy. In 1998, the regional sales volume grew by about 4 per cent, while the Dutch economy grew by 3.75 per cent. This relatively high rate is driven by large business investments and the growth of consumption. Regional employment shows an annual growth of about 2 per cent. The economic success concerns not only the region as a whole but also the city of Amsterdam itself. In contrast with previous years, city employment rose considerably by about 5,000 jobs each year. The city centre has appeared to reap the fruits of increased opening hours on Sundays and the rising number of visitors, resulting in more jobs in consumer services (Economic Development Department, 1997). Amsterdam, and the city centre in particular, can be characterised by a good quality of life, which is illustrated by the increasing popularity of living in the city centre. The city offers a generous supply of cultural facilities, such as museums, theatres and many other institutions.

Table 3.1 Employment in the Amsterdam Region (1995)

	Sector	Share
Primary sector:	Agriculture	1%
Secondary sector:	Manufacturing/construction	15%
Tertiary sector:	Trade	19%
	Health care/social services	20%
	Commercial services	16%
	Transport/warehousing	11%
	Financial services	7%
	Governmental	6%
	Hotels and restaurants	5%

Source: City of Amsterdam, 1995.

One of the engines of the regional economy is tourism. Although the exact size of the tourist sector of Amsterdam has not yet been quantified, it is obvious that tourism generates much employment, illustrated by the share of employment in hotels and restaurants (see Table 3.2). Furthermore, tourism creates jobs in commercial services (congress centres and exhibitions halls) and in retail trade. In comparison with other European cities, the competitiveness of Amsterdam as a tourist city is good. In 1997, the city registered more than 6.755 million overnight stays. In a 1996 comparative study Amsterdam ranked fourth in the top 10 of most-visited cities in Europe. In 1991, Amsterdam occupied the sixth position, as depicted in Table 3.3. Tourism in Amsterdam will be discussed in more detail in the next sections.

Table 3.2 Economic development of the Amsterdam region and the Netherlands; relative annual changes

	1995	1996	1997*	1998*
Gross national/regional product				
Netherlands	2.30	3.30	3.25	3.75
Amsterdam	3.00	2.80	3.50	4.00
Employment (person-years)				
Netherlands	1.50	1.80	2.00	2.25
Amsterdam	2.10	1.90	2.00	2.00

* Estimates.

Sources: SEO, 1997; CPB 1997, adapted by the Economic Development Department (City of Amsterdam) and the authors.

Table 3.3 Top 10 of most visited cities in 1991 and 1996

Rank	Registered overnight stays (1991)		Registered overnight stays (1996)	
1	London*	65,900,000	London*	47,534,000
2	Paris	18,786,000	Paris	18,037,000
3	Rome	6,966,000	Rome	9,314,000
4	Vienna	6,130,000	Amsterdam	6,320,000
5	Nice*	5,351,000	Vienna	6,281,000
6	Amsterdam	5,338,500	Prague	5,950,000
7	Dublin*	4,969,000	Dublin*	5,776,000
8	Edinburgh	4,500,000	Nice*	5,282,000
9	Budapest	3,977,000	Barcelona	4,242,000
10	Prague	3,705,000	Madrid	4,089,000

* Uncertainty about comparability exists.

Source: The Federation of European Tourist Offices (FECTO) and Vienna University; adapted by VVV Amsterdam and the authors.

3 Characteristics and Development of Tourism

This section discusses characteristics and development of tourism in general and of tourism in Amsterdam in particular.

3.1 Characteristics

Visitors are the consumers of tourist products. According to the World Tourism Organisation, the notion of visitor refers to any person travelling to a place other than that of his usual environment for less than 12 consecutive months and whose main purpose of travel is other than the exercise of an activity remunerated from within the place visited (WTO, 1995).

The term 'visitors' comprises both tourists (overnight visitors) and one-day visitors (excursionists). It is important to distinguish visitors who make their trip for the purpose of leisure from those whose main purpose is business. Furthermore, visitors can be classified by their country of residence. Most long-distance visitors are overnight visitors as well.

Cities supply a multitude of functions and facilities for business and leisure tourists. As residents and commuters use the supplied facilities as well, it is important to assess the share of tourists in the total number of users. Berg, Borg and Meer (1995) distinguish *primary tourist products*

In 1996, more than three million overnight visitors came to Amsterdam, which resulted in 6.3 million registered overnight stays. From a survey among 6,000 visitors to Amsterdam it can be concluded that this number is only half of the total number of tourists. The group of nonregistered overnight visitors consists of people who stay the night with family or friends. The number of excursionists is estimated at about 9 million a year. More than 30 per cent of the overnight visitors are travelling for business. The main markets for registered overnight stays in Amsterdam are Great Britain (17.9 per cent), the United States (12.7 per cent) and Germany (9.9 per cent) (VVV, 1998).

from *complementary tourist products*. Primary tourist products are the main purposes of the visit, such as (cultural) attractions and events. Examples of complementary tourist products are hotels, restaurants, conference centres and exhibition halls. Complementary tourist products do not themselves draw visitors, but contribute to the attractiveness of a city's primary tourist products.

Berg et al. (1995) make clear that the attractiveness of a tourist city depends not only on the tourist product but also on the city's image and the internal and external accessibility of the tourist product. The internal accessibility of the tourist product relates to the extent to which tourists can find their way within the city, while external accessibility concerns the international accessibility by car, train and aeroplane. Tourist attractiveness can be influenced by national and local governments (and intermediary organisations) and the tourist industry itself.

3.2 Developments

In Europe, the growth of the tourist market has been explosive since the early 1960s. Rising incomes, an increase of leisure time and more mobility amongst the population have triggered the massification of tourism in the 1960s and 1970s. From the 1980s onwards, tourism continued to grow. The number of arrivals in Europe and the world as a whole increased by, respectively, 60 and almost 80 per cent between 1985 and 1995. Meanwhile tourism became more and more important for the economy, which can be illustrated by the rising international tourist receipts (see Table 3.4). Tourism is expected to be *the* growth sector of the twenty-first century. The expectation is based mainly on the development of the Asian market, and more in particular the Chinese and Indian market.

Table 3.4 International tourist receipts (in US$ million)

	1993	1994		1995		1996	
World	321,124	352,645	+9.8%	401,475	+13.8%	433,863	+8.1%
Europe	162,854	178,412	+9.5%	206,840	+15.9%	219,670	+6.2%
Netherlands	4,690	5,612	+19.7%	5,762	+2.7%	6,256	+8.6%

Source: World Tourism Organisation, 1996.

Next to quantitative changes, qualitative changes can be observed in the tourist industry. Most of these changes are in favour of city tourism (Borg and Gotti, 1995):

- people go on holiday more often;
- holidays get shorter (and short trips have short-distance destinations);
- typical tourist seasons disappear;
- demand for culturally active vacations rises (owing to an increased level of education).

Because tourism generates much employment and qualitative changes in the tourist market are in favour of city tourism, many cities try to attract visitors, which results in much competition between them. However, current success is no longer a guarantee for future success. To remain attractive to visitors, cities have to maintain or improve the quality, quantity, accessibility and image of the city's tourist facilities.

The other side of the coin is that visitor attractiveness can in turn be diminished by the growth of the tourist cluster. If the demand for tourism rises to a level where the quality and accessibility of attractions are compromised to the detriment of society and tourism itself, then growth is no longer acceptable. The maximum limit to the development of tourism is generally known as the carrying capacity (Borg and Gotti, 1995).

4 The Tourist Cluster of Amsterdam: the Main Actors

The tourist cluster of Amsterdam consists of many actors. The main actors can be classified under three headings: firms; research and education institutes; and marketing and promotion organisations. Because government can also influence the attractiveness of the tourist products, public policies regarding tourism will be discussed as well.

In Amsterdam, the number of registered overnight stays has increased by 12 per cent between 1990 and 1997. The average annual growth is 1.7 per cent, but with some serious ups and downs (as depicted in Table 3.5). The Gulf Crisis in 1991 resulted in a decrease of 5.9 per cent, while in 1993 the number of visitors declined by more than 12 per cent. From 1993 onwards the number of overnight stays has increased by almost a quarter at an average growth rate of 6 per cent. In the same period, the occupation rate of hotels increased from 63 per cent to almost 80 per cent. A large proportion of this growth is generated by the Asian market. In 1996, the number of visitors from Japan increased by 18.6 per cent to almost 60,000. The number of overnight stays by people from southwest and north Asian countries grew by more than 30 per cent. Currently, the Asian crisis has a negative effect on the quantity of visitors from Asia.

Table 3.5 Registered overnight stays in Amsterdam

Year	Registered overnight stays	Relative change
1990	6,030,000	
1991	5,679,000	-5.9%
1992	6,191,000	+9.0%
1993	5,427,000	-12.3%
1994	5,954,000	9.7%
1995	6,127,000	2.9%
1996	6,433,000	5.0%
1997	6,755,000	5.0%

Source: VVV Amsterdam, 1998.

4.1 Firms

In Amsterdam, many firms are directly or indirectly dependent on tourists. These firms include not only producers of the primary tourist product – attractions and events – but also producers of complementary goods, such as the hotel and catering industry, conference centres and exhibition halls, shops and tour operators.

Attractions and events Among the important attractions in Amsterdam are the canal touring boats (about 2.2 million visitors), the Rijksmuseum (1,302,887) and the Van Gogh Museum (1,144,128) (VVV, 1997). In Amsterdam cultural

Table 3.6 Cultural attractions in the city centre and Amsterdam

	City centre	Other Amsterdam	Share of city centre
(Music) theatres	80	51	61%
Cinemas	11	3	79%
Museums	34	11	76%
Total	125	65	65%

Source: DRO and Amsterdams Uitburo, 1997.

Table 3.7 Hotels in Amsterdam (1997)

Category	Hotels	Rooms	Guests	Nights	Occupation rate	Average length of stay (days)
Five-star	16	4,080	855,000	1,820,000	0.79	2.13
Four-star	27	3,610	725,000	1,545,000	0.74	2.13
Three-star	66	3,880	715,000	1,625,000	0.76	2.27
Other	195	3,890	760,000	1,750,000	0.65	2.34
Total	304	15,460	3,055,000	6,740,000	0.74	2.20

Source: City of Amsterdam/O+S, 1998.

and leisure pursuits can be easily combined. On the one hand Amsterdam offers the cultural tourist a range of more than 40 museums, on the other hand the city attracts people to its well-known red-light district. The majority of the attractions are concentrated in the city centre and around the Museumplein (Rijksmuseum and Van Gogh Museum). Many events take place in Amsterdam, most of which are annual events (such as the Uitmarkt with about 500,000 visitors). Every five years, Sail Amsterdam is organised, which shows an impressive number of sailboats and attracts more than four million viewers. Many events (exhibitions) take place in the RAI (see below).

Hotel and catering industry Visitors to Amsterdam can make their choice from 320 hotels in which to stay the night. With 190 hotels, the city centre has the largest concentration of hotels. The majority of the other hotels can be found south of the city centre.

Within the various quality classes, a shift can be observed. The proportion of nights in four-star and five-star hotels has increased from 43 per cent in 1995 to 53 per cent in 1996 (Dienst Binnenstad Amsterdam, 1998). With an occupation rate of almost 80 per cent, the demand for hotel rooms is often greater than the supply, especially in the traditional tourist seasons (summer and other holidays), driving up hotel prices. In a policy document, the city sets out its aims of creating an extra supply of about 5,000 hotel rooms before 2003 (City of Amsterdam, 1998). Because the city centre does not offer enough space, hotels must be built elsewhere.

The catering industry is also well represented in the city centre with 1,460 pubs and restaurants (ibid.). New developments in the field of pubs and restaurants can be observed at the Rembrandtplein. From the 1970s onwards, this square slowly lost its dignity. During the 1990s, however, it developed itself into a new and trendsetting entertainment centre. Amsterdam nightlife has shifted increasingly from the Leidseplein to the Rembrandtplein.

Conference centres and exhibition halls An important player in the tourist cluster, especially with regard to the business travellers, is the Amsterdam RAI. This international exhibition and congress centre receives about two million visitors a year. The RAI is not only offers accommodation but is also an organisation, arrangomg 140 exhibitions all over the world. The complex is located along the South Axis and has its own railway station. It consists of 10 halls with approximately 100,000 m^2 meeting and exhibition space. The events can be classified into international conventions (more than 40), international trade fairs and exhibitions (more than 70) and national conferences and meetings. On average, visitors to conferences in the RAI stay four to five days in Amsterdam and spend 2,000 guilders each. International events especially have an enormous impact on the city. Next to the RAI, there are also 42 conference hotels in Amsterdam suitable for conferences with fewer than 200 participants.

Conferences and exhibitions are increasingly combined, particularly at medical congresses, which have relatively many participants. Other important developments are concentration and deconcentration. Owing to the realisation of a single European market, the importance of European associations is exceeding that of national associations, resulting in conferences with more participants (increase in scale). Moreover, one can observe a rationalisation concerning the choice of a conference city. On the other hand, new specialisations on the European level lead to deconcentration. Most of these conferences take place in hotels.

To safeguard its current position in Europe, the RAI is expanding its complex by more than 25,000 m^2. Moreover, the complex will be modernised to meet new standards. The expansion mainly concerns exhibition space but includes a five-star hotel, which will be integrated into the complex. The international accessibility of the exhibition and conference centre will be improved by a connection to the high-speed train network. These trains will stop at South Amsterdam, but not at Amsterdam RAI. RAI visitors will have to travel by underground (the new North-Southline,[2] which will be completed in 2006) to the RAI's underground station (also yet to be constructed).

Shops Amsterdam is not specifically known as a shopping city, in contrast with, for instance, Paris, London and Düsseldorf. 'Fun shopping', however, is a growing business in Amsterdam, and in the city centre in particular. A large proportion of the 2,000 shops in the city centre can be described as 'fun shops'. These shops are concentrated in several areas. The most important and well-known area comprises the Leidsestraat and the Kalverstraat, with branches of international chains of shops. Many tourists pay a visit to the Bijenkorf, one of the main players in the field. The shops in this part of Amsterdam are open every Sunday, drawing many one-day visitors to Amsterdam. There are some other shopping centres in Amsterdam, but most of them are unknown to the majority of tourists.

Tour operators Tour operators in Amsterdam, such as Holland International, sell packages to tourists, including a trip to Amsterdam, a few nights at a hotel, and excursions to interesting attractions in and around Amsterdam. They play an important role in the tourist cluster, because they have contacts with consumers as well as suppliers. Using these contacts, they can create packages to anticipate the demands of the 'zap consumer'.

Foreign tour operators can buy packages from tour operators in Amsterdam, including hotels, or contact hotels in Amsterdam to take rooms in allotment. Allotments – reservations of a specific room on a specific day – give tour operators the right to book a room up to one week before the day in question. Because of the success of Amsterdam, tour operators in Amsterdam and in foreign countries are having difficulties in getting allotments, especially during tourist seasons (holidays). Some tour operators have already decreased the number of pages about Amsterdam in their brochures in favour of other cities.

Research and education institutes Knowledge about tourism, generated by research and education institutes, can improve the quality of the tourist product. Three universities are located in Amsterdam: the University of Amsterdam (with more than 20,000 students), the Free University of Amsterdam (with almost 15,000 students) and the Catholic Theological University of Amsterdam (with 80 students). The University of Amsterdam and the theological university are located in the city centre, while the Free University has its buildings south of the city centre. In Amsterdam, research into tourism is mainly carried out by the VVV (see below).

Institutes for further education and training in a city can contribute to the quality of hotels and restaurants. An example of such an institute is the 'Hotelschool' in Amsterdam, which trains professionals for the hotel and catering industry. In general, Amsterdam is characterised by a high density of research and education institutes. These institutes generate a local demand for conferences and exhibitions, and thus contribute to the development of business tourism. Amsterdam offers a great variety of arts and cultural educational opportunities and the city's conservatoire is considered the best in the Netherlands.

4.2 Marketing and Promotion Organisations

One of the main players in the marketing and promotion of the tourist cluster of Amsterdam is the Tourist Board (VVV Amsterdam). The members of this association help to develop tourism in Amsterdam and its region (VVV, 1997). The VVV aims at providing information to tourists and the Amsterdam population, promoting Amsterdam as an attractive city, safeguarding and improving the quality of the tourist product and advising tourism-related public and private actors. To reach these goals, the VVV wishes to cooperate with private actors to create an integrated tourist product.

The VVV Amsterdam is due to be reorganised. A holding company will be established, consisting of four corporations. A knowledge centre will be the core business of the organisation. The three remaining corporations willbe responsible for information distribution (front offices), reservations and consultancy. Each of the four profit centres mentioned will have its own director, pursue its own strategy and be responsible for its own results. The implication is that they may have to compete with other actors in the tourist cluster. The booking agency, for instance, may have to vie with tour operators.

Other relevant marketing and promotion organisations are Amsterdam Congresbureau (ACB), the Chamber of Commerce, Amsterdam Promotion (AMPRO), KLM and NBT. ACB is an initiative of congress hotels and congress centres in cooperation with local government. It aims at attracting congresses to Amsterdam. Amsterdam Promotion is a joint venture of public and private actors that attempts to promote Amsterdam to (inter)national businesses. KLM is the largest Dutch airline company and actively promotes the Netherlands and Amsterdam as a place to visit. The NBT (Nederlandse

Bureau voor Toerisme; Dutch Office for Tourism) is also involved in the promotion of Amsterdam.

4.3 Public Policies Concerning the Tourist Cluster

Tourism is one of the spearhead industries in the policy of the city of Amsterdam. The city government acknowledges the importance of tourism for the regional economy and accordingly, tourism is actively stimulated. Reasoning from a market philosophy, the city has adopted a favourable policy towards the tourist cluster. While the VVV is mainly responsible for promoting the city as a tourist destination, the city of Amsterdam, and its Economic Development Department (EDD) in particular, stimulate tourist product development. Product development refers not only to the creation of new attractions (such as NewMetropolis, an interactive science museum) but also to the maintenance and extension of existing attractions (such as the Rijksmuseum and the Van Gogh Museum). An interesting example of product development is the initiative of the Nieuwe Kerk (an important centre for arts and exhibitions) to develop an 'annex' of the well-known Hermitage of St Petersburg in Amsterdam.

Several sources of money from EDD to initiatives and institutes with regard to tourism in the city can be identified. Half of the VVV's budget is financed by tourist taxes. The budgets of Amsterdam Promotion and the Amsterdam Congresbureau are partly financed by public means as well. Moreover, EDD grants subsidies to organisers of new events in the city. Other divisions of local government also pay attention to tourism. For instance, the municipal transport department (which is responsible for public transport) offers tourists multilingual signs and maps.

The city of Amsterdam is aware that there are negative elements to tourism. Some parts of the city centre are so dominated by tourism that businesses and citizens complain about it. Therefore, the city of Amsterdam wants to spread tourism in two ways. For one thing, tourists should be distributed in space. Other parts of the city, such as the eastern part of the city centre, are interesting to tourists as well. In the perception of the average international tourist, 'The Netherlands' are identical to a small part of Amsterdam (the western part of the city centre). For another, tourists should be distributed in time. Many visitors come to the city during holidays. A more equal distribution of tourists in time and space would reduce the pressure on the historic city centre. Public policies with regard to tourism are in accordance with a policy plan for the city centre, which stresses the importance of maintaining a mix of functions in the city centre (City of Amsterdam, 1993).

Spread of tourism in space is partly influenced by the distribution of hotels across the region. The city discourages hotels from settling in some parts of the city centre, but it is not actually forbidden to build a new hotel in these areas. That would be contrary to the market philosophy, mentioned above. Moreover, the city supports the realisation of 5,000 extra hotel beds. The city does support initiatives to attract tourists to areas outside the traditional tourist zones, such as an initiative from attractions located in the eastern part of the city centre.

To spread tourism across the region, the city cooperates with crowd-pullers in the region, such as Marken, Volendam, the Zaanse Schans, the Keukenhof and also Delft to stimulate tourists to spend the night close to these attraction instead of in the city centre. It seems to be difficult to persuade visitors to book a hotel outside the city (centre). Tourists are focused on the city (centre) of Amsterdam. Since 1995, the city of Amsterdam and the VVV have adopted a wider vision on tourism and broadened the focus of the communication policy to incorporate other attractions outside Amsterdam.

5 Interaction and Dynamics in the Cluster

Tourism in Amsterdam is booming. In a competitive environment like the tourist industry, success may come to an end overnight. Therefore, cooperation and interaction between the different actors in the cluster is decisive, not only to fulfil the wishes of the 'zap consumer' but also to create an integrated tourist product. Strategic networks are likely to benefit all actors, because a more attractive tourist product (or a better image) will draw more visitors to the city. In this chapter, the main strategic interactions between the various parties will be described. In figure 5.1, these interactions are schematically presented. The section numbers correspond to the numbers in the figure.

5.1 Interfirm Relations

In spite of a large degree of competition between firms in the tourist cluster, actors do cooperate. In many cases cooperation is based on informal contacts. However, there are some formal strategic networks as well, based on the recognition of common interests.

Shops are united in the VAC (Vereniging Amsterdam City; Association Amsterdam City). This association protects the interests of the various shops

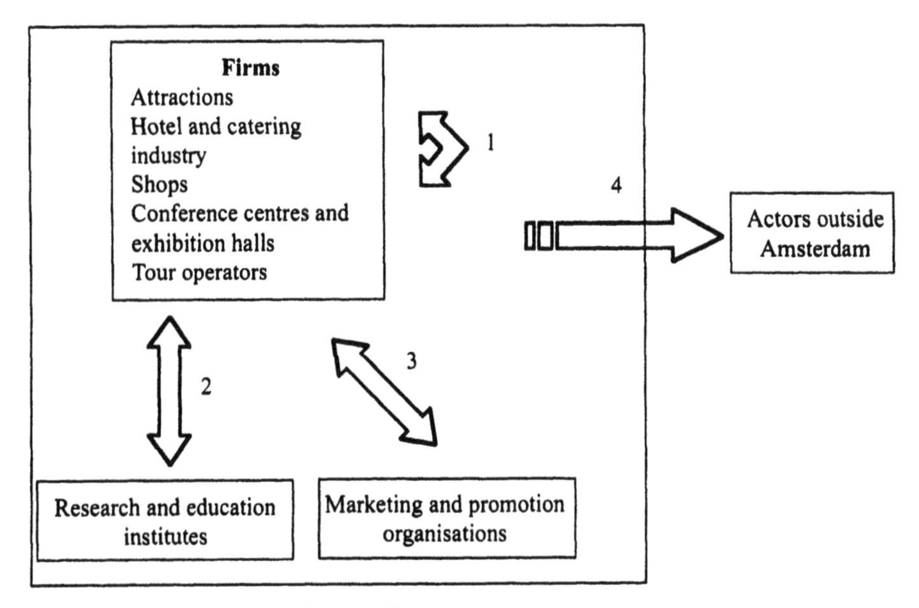

Figure 3.3 Relations in the cluster

in Amsterdam's city centre. For instance, the VAC tries to minimise the negative effects of the construction of the North-Southline. Furthermore, the Association wants to make the whole range of shops more visible to visitors. At present, many tourists visit only the Kalverstraat-Nieuwendijk area, neglecting other pleasant shopping areas, such as the Jordaan. Together with the VVV, the VAC wants to lay out some shopping routes to show tourists the diversity of shops. Cultural attractions are also represented in the VAC. Both shops and attractions can help to spread tourism in time by an extension of opening hours.

Cooperation between attractions tends to be incidental. In the eastern part of the city centre, however, a cooperative project, called 'Plantage aan het water' ('plantation along the water'), has been set up by 11 cultural attractions, to improve the tourist climate of this area. To these attractions belong such main actors as the Scheepvaartmuseum (maritime museum) and Artis (the local Zoo). Together they try to develop new products, such as a joint event or joint ticketing. The 'Scheepvaartmuseum' (maritime museum), Madam Tussaud Amsterdam and Artis Zoo, conscious of the importance of product development, have created a common package. Moreover, some attractions in this part of the city want to join forces for general communication, for instance by offering one common brochure.

In the hotel business, informal networks are used to discuss developments with relevance for the hotel industry and to evolve a common vision on the future of hotels in Amsterdam. Cooperation between hotels is aimed at protecting common interests and strengthening the position towards tour operators. Moreover, hotels are united in Horeca Nederland (Hotel and Catering Industry Netherlands), an employers' association. At meetings, employers debate local policies or the expected effects of a specific event. In the context of EURO 2000, for instance, the hotels want to make a common proposal to the UEFA about sale conditions.

Tour operators have many contacts with hotels and attractions to create all-in packages. Their role in the tourist cluster, however, is threatened by new media, such as the Internet. Krasnapolsky, one of the five-star hotels in Amsterdam, has already opened an Internet site for booking rooms.[3]

The RAI has an intensive relation with almost all hotels in the region. RAI Hotel Service offers visitors to the RAI a complete service, including advice and reservation. Many visitors to the RAI take advantage of the opportunity to see one of the various museums and other attractions in Amsterdam, spending more money than does the average tourist.

Since hotels and attractions in Amsterdam are conscious of their common interests, they meet in formal and informal ways. They do not, however, create joint packages. The occupation rate of hotels is already very high and most attractions have no complaints about the number of visitors. As an example: the concert hall is often sold out, making it very difficult for hotels to put together special packages for their guests. The relationship between hotels and attractions is limited to the provision of information to the hotel guests. This is remarkable, since events generate a lot of extra hotel guests.

5.2 Relations between Firms and Research and Education Institutes

Firms in the tourist cluster can benefit from the presence of research and education institutes. Although tourism, like the financial and cultural clusters, generates much employment in Amsterdam, relatively few education programmes for tourism are available in the city. The University of Amsterdam does offer a lecture on the science of leisure, but has no strategic relations with the tourist cluster of Amsterdam, apart from being, like other education institutes, represented in the Vereniging Amsterdam City.

It is certainly true that many jobs in tourism do not require an academic degree, but the tourist industry would benefit from the presence of vocational training institutes. In Amsterdam, the number of professional training courses for tourism is surprisingly low, considering the impact of tourism in the city. The relations between the tourist industry of Amsterdam and the education institutes of all levels are underdeveloped. In view of the present professionalisation and upgrading of tourism, it is certainly advisable to strengthen these relations in order to maintain or improve the city's current position in the competitive tourist industry. Moreover, education institutes and universities in particular, are important players in the field of conferences. In that sense, they contribute to the development of business tourism.

5.3 Relations between Firms and Marketing and Promotion Organisations

In 1996, the VVV Amsterdam formulated a strategic policy vision oriented to 2005. One of the objectives of this vision was to strengthen the relations with firms inside and outside the tourist cluster, in order to create an integrated tourist product. Currently, cooperation with the private sector and in particular the cultural institutions is rather good. However, strategic interactions with the shops in Amsterdam could be improved. Accordingly, the VVV works together with the VAC to lay out shopping routes, in order to make the complete range of shops more visible to the tourists.

The reorganisation of the VVV has increased the possibilities for cooperation. Every corporation will have its own partnerships. Front offices play an important role in the provision of information. The VVV can help to increase the visibility of packages created by attractions and/or hotels. Furthermore, tour operators use the VVV packages to build their own. One of these packages is 'Amsterdam, an art adventure', which covers a number of cultural events in the off-season period. On the other hand the VVV booking agency also sells packages itself, competing with tour operators. Finally, the consultancy corporation can advise actors in the tourist cluster, making use of the knowledge that has been gathered by the knowledge centre.

Although Amsterdam Promotion's main goal is to attract businesses to Amsterdam, this organisation certainly plays a role in visitor attraction as well. AMPRO is a good illustration of the advantages of cooperation between public and private actors. This organisation has a clear vision, shared by the actors involved, on the way Amsterdam should be promoted abroad. The same holds for the Amsterdam Congresbureau, which takes care of promoting Amsterdam as a congress city.

5.4 Relations between the Tourist Cluster and Actors outside Amsterdam

As a member of the Federation of European Cities' Tourist Offices (FECTO), the VVV of Amsterdam shows willingness to cooperate with other cities in Europe. The main driving force is the fact that people from outside Europe usually combine visits to several cities in Europe. Joint promotion and marketing efforts are necessary to attract people to a specific area in Europe. FECTO has divided Europe into so-called association areas. Cities in an association area are often combined in one tour by long-distance visitors. Amsterdam is part of the same area as Paris, Brussels and Berlin. Cooperation between these cities is still in its infancy, however, especially with regard to joint product development and communication.

In the national context, the VVV cooperates with the NBT and KLM, the largest airline company, with Schiphol as its home base. All being involved in the marketing of Amsterdam as a tourist city, these three organisations have set up NBT Amsterdam, which coordinates the promotion of Amsterdam abroad. VVV Amsterdam commissions NBT to carry out the promotion of Amsterdam in foreign countries. With this division of labour, the tourist cluster of Amsterdam can gain optimal profit from the market knowledge of NBT establishments abroad.

Many actors in the tourist cluster agree that the spread of tourism in space should be stimulated. There is not enough space in the city centre to accommodate all tourists. Visitors to Amsterdam, however, are unwilling to sleep in a hotel outside the city. They refuse to spend the night in Rotterdam (only one hour from Amsterdam), and even in suburbs of Amsterdam, such as Amstelveen.

6 Confrontation with the Framework

The tourist cluster of Amsterdam cannot be isolated from its social, economic and spatial context. Therefore, this section presents a confrontation with the framework of reference, explained in Chapter Two, to analyse the factors that determine growth, such as the spatial and economic context, organising capacity and cluster development.

6.1 Spatial and Economic Context

The tourist industry is highly dependent on macroeconomic developments. The growth of the economic cluster is to a large extent determined by factors that cannot be influenced by the actors in the city. Recessions, wars and currency fluctuations have a great impact on the number of visitors. In 1998,

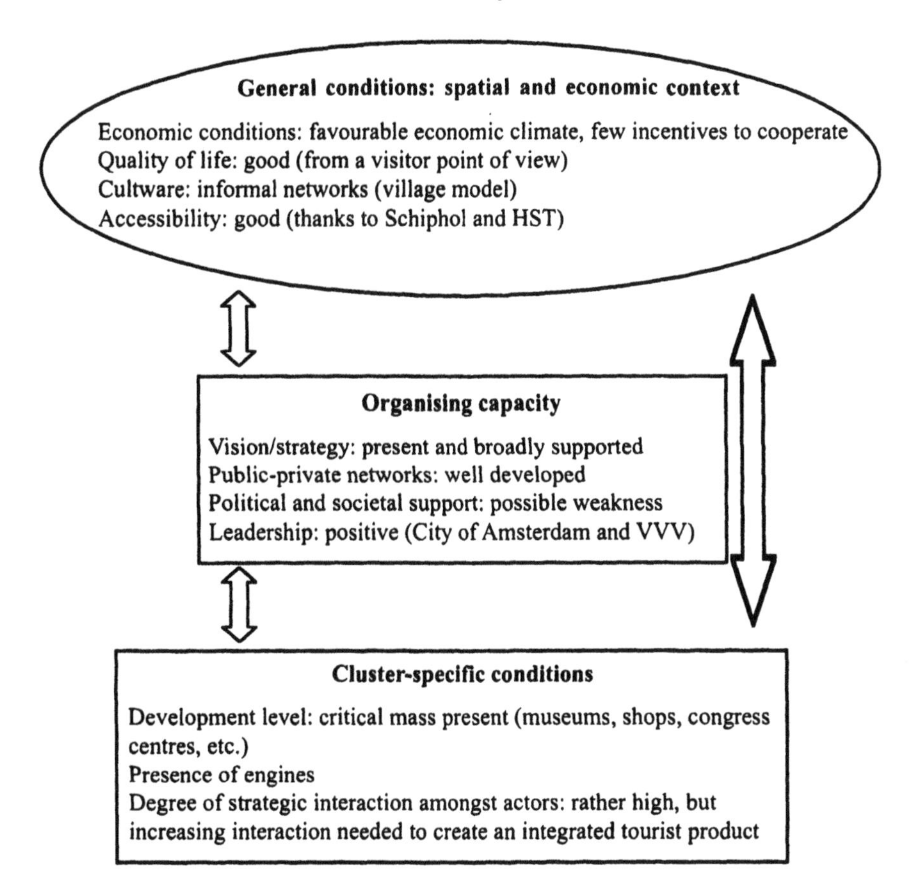

Figure 3.4 Framework of reference

the number of Irish visitors increased enormously, not because Amsterdam had become more attractive to Irish people, but because the Irish economy was flourishing. In spite of the Asian crisis, Amsterdam has no complaints about the number of visitors. Occupation rates are continuously rising and some hotels in the city centre (near the Dam) even show an occupation rate of 90 per cent! In some cases, however, success makes hotels and attractions in Amsterdam passive with regard to cooperation. The development that the number of pages in brochures of foreign tour operators is declining is a serious threat for the future of the tourist cluster of Amsterdam.

On the other hand, informal networks are easily created in a 'village' like Amsterdam. The existence of informal networks is a result of historical factors. The fact that everybody knows one another leads to 'gentlemen's agreements' that raise the quality of the tourist cluster.

In contrast with other clusters, quality of life concerning tourism also has to be determined from a visitor point of view in conjunction with an employee point of view. The quality of life in Amsterdam is good, as is shown by the fact that living in the city centre of Amsterdam is increasingly popular, resulting in rising prices. The wide range of (cultural) facilities attracts high-income groups to the city centre.

Another important determinant of the spatial context is accessibility. Internal and external accessibility determine the attractiveness of the tourist product. The city of Amsterdam is very accessible thanks to the proximity of Schiphol Airport and the connection to the international network of high-speed trains. To safeguard future growth, it is very important to maintain the present reputation of Schiphol Airport. Discussions about so-called 'slots'[4] can do serious harm to this reputation. For shorter distances, high-speed trains (HSTs) can function as substitutes for aeroplanes. Therefore, it is important that HST-travellers can easily reach the city centre, where tourist attractions and hotels are concentrated. Furthermore, business tourists will appreciate a fast and pleasant connection with Amsterdam RAI. The high-speed train does not stop at the RAI itself but at Amsterdam South, the city's top business location. Visitors to conferences and exhibitions have to change trains. In Cologne, for instance, the HST-station is integrated into Cologne's conference centre and exhibition hall.

6.2 Cluster Development

The previous sections will have made clear that the tourist cluster of Amsterdam is well developed. Given the number of attractions, there is no doubt about the presence of critical mass. The quality of the actors is good and will be better. An example is Amsterdam RAI, which will raise its quality by modernisation and expansion. Hotels are using the present upswing in tourism to upgrade their facilities, anticipating future demands of visitors. New shopping malls such as the Kalvertoren in the Kalverstraat contribute to the attractiveness of Amsterdam as a shopping city.

A weak point in the tourist cluster of Amsterdam is the scarcity of hotels, but plans have been made to build new hotels, mainly around the city centre. The absence of group entertainment shows is another weak point in the supply of tourist products. Private companies are reluctant to invest in such shows, because they think there is not enough demand for them. However, international shows (in English, or shows that do not depend on language) seem to be viable, given the number of foreign tourists.

The interaction between the actors is fair, thanks to the village culture in Amsterdam. Nevertheless, relations could be strengthened. Cooperation implies more than talking with each other. Firms in the tourist cluster could pay more attention to the creation of packages to comply with the wishes of the 'zap consumer'. A higher degree of cooperation between the actors in the tourist cluster could produce a more integrated tourist product that could benefit all actors. Furthermore, the most important education are not in the Amsterdam region. Cooperation does not refer only to product development but also to marketing and promotion. A common promotion policy is important, because image is decisive for the attractiveness of cities. Cooperation between VVV Amsterdam, NBT and KLM has already contributed to a univocal marketing policy for Amsterdam.

6.3 Organising Capacity

The growth potential of a cluster is partly determined by its organising capacity. The organising capacity depends on the presence of a vision and a strategy, the quality of public-private networks, the level of political and societal support for the development of the cluster, and leadership.

The VVV and the City of Amsterdam have a clear vision of the development of the tourist cluster. In this vision, the number of visitors to Amsterdam is allowed to grow under a number of conditions. Growth should be limited to 4 per cent a year to avoid too much pressure on the city. Furthermore, the city of Amsterdam aims at realising this growth in off-season periods (spread in time) and in areas outside the city centre of Amsterdam (spread in space). The VVV and other marketing and promotion actors act in accordance with this vision. Moreover, many firms in the tourist cluster support the vision and strategy of the local government and the VVV. The distribution of tourists in time seems to have been realised to some extent, for seasonal influences have almost disappeared. To spread tourism in space, however, seems to be more difficult. City boroughs other than the city centre are willing to attract visitors, but most visitors focus on Amsterdam's city centre.

Public and private networks are well developed. Informal relations have proved to be the foundation of several strategic partnerships, such as the VVV, the Vereniging Amsterdam City, Amsterdam Congresbureau and Amsterdam Promotion. However, from a regional point of view, networks could be improved. The VVV of Amsterdam already cooperates with the provincial VVV[5] and the NBT, but more strategic interaction with other cities in the

Amsterdam region could promote the spread of tourism in space and the creation of an integrated regional tourist product.

One possible weakness of the organising capacity of the tourist cluster in Amsterdam refers to political and societal support. According to the policy plan for the city centre, the city of Amsterdam aims at maintaining a diversified economic structure in this part of the town (City of Amsterdam, 1993). Tourism is just one of the economic functions of the city centre, next to other economic sectors (such as finance) and living. Too much pressure on the city centre caused by too much growth of tourism can result in an exodus of businesses and inhabitants (crowding out). Already, citizens of Amsterdam and people who work in Amsterdam complain about the lack of parking places because of the massive throng of tourists to the city. Because tourists make use of the same facilities as the local population and employees, such as public transport, the growth of tourism should be limited. Furthermore, growth should be realised during off-seasons and in other parts of the town than the city centre. In order to maintain societal support, the city of Amsterdam and the VVV also need to communicate the importance of the tourist industry to the citizens and the private sector.

Finally, leadership is a positive element of the organising capacity of Amsterdam's tourist cluster. The city of Amsterdam and the VVV have been able to create cooperativeness by formulating an integrated vision and strategy, aimed at raising the quality of the tourist cluster.

7 Conclusions and Perspectives

In competition with other cities around the world, Amsterdam has good potential to maintain its present position as one of the most visited cities in Europe. The unique historic city centre and the impressive supply of museums and other attractions contribute to the appeal of Amsterdam as tourist city. However, many other European cities invest heavily in the development of their tourist products. Therefore, one of the most important recommendations is not to sit back and wait. To secure its current status, Amsterdam needs constantly to raise the quality of its primary and complementary tourist products. In Amsterdam, local government and various firms (hotels, attractions and congress centres) acknowledge the necessity of constant vigilance with regard to the quality and quantity of tourist products.

To fulfil the wishes of the twenty-first century consumer, who will put together his or her own packages, actors in the tourist cluster must cooperate.

On the whole the conclusion is that the degree of interaction in the tourist cluster of Amsterdam is already rather high, thanks to many informal relations amongst the actors. Strategic interaction with respect to joint product development could be enhanced. High occupation rates and positive visitor statistics have weakened the incentives for hotels and attractions in Amsterdam to develop products together. At present, the economic climate is favourable, but because the number of visitors largely depends on changes in the world economy and consumer preferences, it is advisable to be prepared for a less favourable economic climate.

A number of challenges emerge from the analysis of the tourist cluster of Amsterdam. First of all, the intensity of strategic relations between education institutes of all levels and firms in the tourist industry could be improved. The professionalisation trend in the tourist industry requires that educational programmes are up to date with the business practice in the cluster. Another possible weakness refers to the societal and political support. Too much tourism in Amsterdam could do away with this support and in the end even harm the tourist industry. It is necessary not only to keep the balance between the various functions of the city centre, but also that between leisure and business tourism.

Too much business tourism might lead to an image of dullness. Thirdly, the city of Amsterdam has to take care of its external accessibility. Currently, external accessibility is rather good thanks to the nearness of Schiphol and a connection to the network of high-speed trains. It is very important to maintain the international status of Schiphol and to optimise the benefits of the HST.

Finally, tourism in Amsterdam has to be placed in a broader perspective. To spread tourism in space, cooperation with municipalities in the region or even cities in the Randstad is needed. Some of these cities, such as The Hague and Rotterdam, could add to the tourist product of Amsterdam, because their tourist products are complementary to those of Amsterdam.

Notes

1 Co-author: drs Alexander Otgaar.
2 The North-Southline will connect the historic city centre with the South Axis.
3 The address of this site is http://www.krasnapolsky.nl/.
4 At Schiphol Airport, the number of flight movements (take-offs and landings) has been limited by the national government. The notion of 'slot' refers to a reservation of a flight movement.
5 VVV Noord-Holland.

References

Amsterdam Airport Schiphol (1998), *Annual Report 1997; Comparison of top 10 West European airports.*

Amsterdam Promotion (AMPRO) (1998), *Amsterdam, Capital of Inspiration.*

Berg, L. van den and Bramezza, I. (1992), *Randstad Holland: Growth and planning of a globally competitive metropolis*, Euricur, Rotterdam.

Berg, L. van den, J. van der Borg and J. van der Meer (1995), *Urban Tourism; Performance and strategies in eight European cities*, Ashgate, Aldershot.

Borg, J. van der en G. Gotti (1995), *Tourism and Cities of Art: The impact of tourism and visitors flow management in Aix-en-Provence, Amsterdam, Bruges, Florence, Oxford, Salzburg and Venice*, UNESCO Regional Office for Science and Technology for Europe, Venetië.

City of Amsterdam (1993), *Beleidsplan Binnenstad.*

City of Amsterdam (1995) *Cijfernota.*

City of Amsterdam (1998), *Hotelbeleid 1999–2003.*

City of Amsterdam/O+S (1998), *Feiten en Cijfers over Amsterdam* (http://www.amsterdam.nl/).

CPB (1997), *Macro-economische Verkenningen 1997.*

Dienst Binnenstad Amsterdam (1998), *Trendrapport Amsterdamse Binnenstad 1997.*

DRO and Amsterdam Uitburo (1997), *NEA-bestand.*

Economic Development Department (EDD) (1997), *De Economie in de Regio Amsterdam 1997*, City of Amsterdam.

Federation of European Cities' Tourist Offices (FECTO) and Vienna University (1996), *City Tourism in Europe.*

SEO (1997), *Algemene Economische Verkenningen 1997.*

VVV Amsterdam (1997), *Jaarverslag 1996.*

VVV Amsterdam (1998), *Strategisch Marketing- en Communicatieplan voor Toeristisch Amsterdam.*

World Tourism Organisation (WTO) (1995), *General Concepts and Definitions*, Maastricht.

World Tourism Organisation (WTO) (1996), *Annual Report.*

Discussion Partners

P. Becker, Holland International, Director.

W. Bijleveld, Scheepvaartmuseum, Director.

D. Elzinga, Amsterdam RAI, Director.

D.K. Freling, Chamber of Commerce (Kamer van Koophandel) Amsterdam.

W. van der Kolk, Vereniging Amsterdam City, Chairman.

J. Moreu, VVV Amsterdam, Director.

W. Vehmeijer, Economic Development Department (Economische Zaken), City of Amsterdam, Director.

F.M. Werners, AMS Hotel Group, Director.

Chapter Four

The Mechatronics Cluster in Eindhoven

1 Introduction

This chapter describes and analyses the highly dynamic mechatronics cluster in the greater Eindhoven region. The aim of the case study is to detect, describe and judge strategic relations between various actors in the cluster (firms, the knowledge infrastructure, local/regional government), and to put the cluster into the perspective of the general economic structure of the region. The case study assesses how organising capacity can be enhanced to further the development of the cluster.

The organisation of this case study is as follows. In section 2, a general profile of the greater Eindhoven region is sketched, as well as the economic development and structure of the region. Section 3 is dedicated to the delimiting and definition of the mechatronics-cluster. In section 4, the most important actors and their activities in the Eindhoven mechatronic cluster are introduced and described. Section 5 analyses the strategic relations between the various actors. In section 6, the mechatronics cluster is put in the framework of reference developed in Chapter Two. Section 7 concludes.

2 Profile of the Region and Economy

2.1 Profile

At the beginning of the twentieth century, the city of Eindhoven was no more than a small agricultural town, with some 5,000 inhabitants. The foundation of Philips Gloeilampen N.V in 1891 marked the beginning of the city's rapid development (Adang and van Oorschot, 1996). Nowadays, Eindhoven is the fifth city of the Netherlands, with 198,000 inhabitants. It forms the centre of the region of South-East Brabant, which is often referred to as the Greater Eindhoven Area, or the 'Eindhoven region' (see Figure 4.1). This region comprises 34 municipalities. Like the city of Eindhoven, the population of this region has grown fast in the last century. Today, it counts some 670,000 inhabitants.

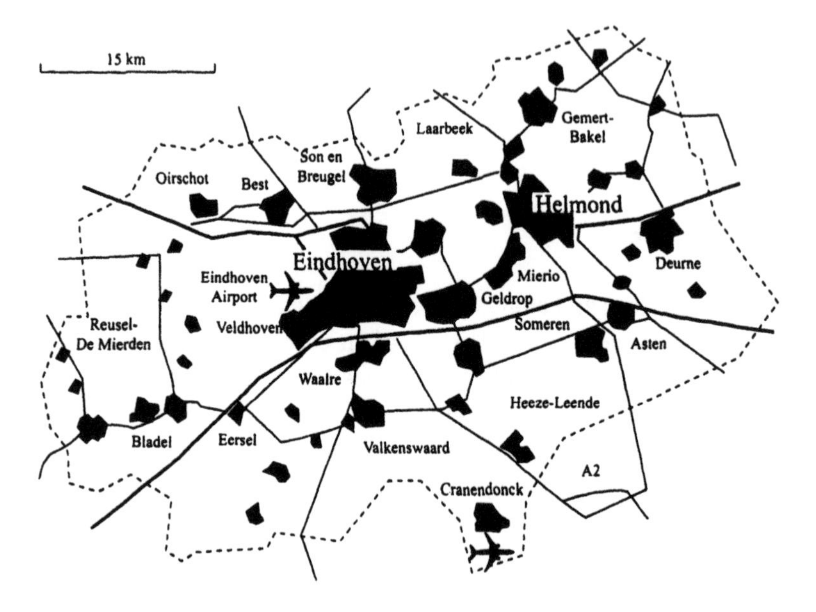

Figure 4.1 The greater Eindhoven region

Source: FPER, 1997.

Table 4.1 Distances from Eindhoven to nearby international cities (in kilometres)

City	Distance
Rotterdam	110
Amsterdam	120
Antwerp	100
Düsseldorf	100
Frankfurt	340
Paris	450

The appearance of the city of Eindhoven is determined by urban design of the twentieth century. In the city centre, buildings and housing estates offer a cross-section of twentieth century (housing) architecture. The city's fast growth and Eindhoven's preoccupation with technology have also left their mark on the city's appearance. Although Eindhoven is the fifth city of the Netherlands, up to the 1980s the urban environment did not live up to that status. The town centre looked chaotic, and apart from the Evoluon there were no special attractions. Since the 1980s, Eindhoven has put much effort in city renewal. A high quality shopping centre (heuvelgalerie) and a concert hall (Muziekcentrum Frits Philips) have been opened, as well as a multi-functional building (the Witte Dame, a restructured large Philips building). Eindhoven is trying hard to bring its urban environment in line with its economic, social and cultural position.

Geographically, the region of Eindhoven is situated at some 100 km west of the Randstad, the cultural and economic 'gravity centre' of the Netherlands. Nevertheless, its location is by no means peripheral, in particular in a European perspective. The region is situated within the rectangle formed by the Randstad, central Germany, the Ruhr Area and the Belgian cities of Brussels and Antwerp. Table 4.1 presents the distances from Eindhoven to some other large cities.

2.2 Economic Development and Structure

The Eindhoven region ranks among the more prosperous in the Netherlands. Incomes per capita are well above the Dutch average. The region of Eindhoven is an important centre of employment. It counts some 307,000 jobs in a population of 670,000.

Table 4.2 Some economic indicators of the Eindhoven region and the Netherlands, 1996

	Eindhoven	Netherlands
Employment growth	2.8	2.1
Investment growth	15.0	3.0
Turnover growth	4.2	3.0
Export growth	7.9	2.1

Source: Polytechnisch Tijdschrift, 1997.

During the last 15 years, the economic fortunes of the region have changed several times. Up to the mid-1980s the region did relatively well, with growth rates higher than the national average. From then on, however, regional employment growth fell below national average. Between 1986 and 1991, the yearly employment growth rates dropped from 5.3 to 0.9 per cent. From 1991 onwards, the situation worsened. On balance, the region started to lose jobs. The regional employment figures continued to drop up to 1994. The main factors behind the economic problems of the region were a reflection of severe difficulties of the two leading industrial firms in the region: Philips (electronics) and DAF (lorry construction). In the early 1980s, Philips employed 35,000 people in the region. This figure had dropped to 21,000 by 1993. In 1993, the DAF company collapsed, involving the loss of another 2,500 jobs. Additionally, the large network of DAF's external suppliers in the region was severely hit. After a few difficult years, the region had recovered strongly by the mid-1990s. As Table 4.2 shows, by many economic indicators the region shows higher growth rates than the Netherlands as a whole.

Along with a general recovery of the economy in the second half of the 1990 in the Netherlands, the drastic restructuring of Philips and DAF at the beginning of the 1990s has generated positive spin-off in the form of newly-created businesses and the outsourcing of activities.

The *economic structure* of the Eindhoven region is dominated by industry, as can be seen in Figure 4.2. But trade, business services and health and welfare services are also large sectors with regard to employment. The share of industry in the regional economic structure (in terms of both employment and added value) is one of the highest in the Netherlands.

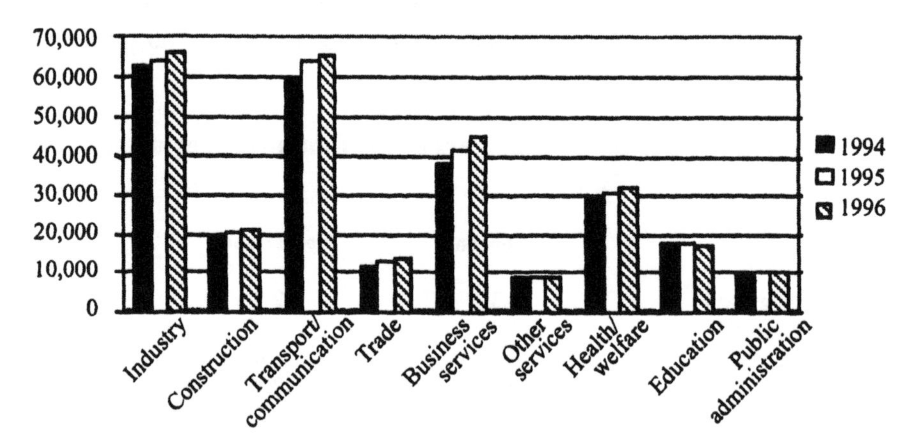

Figure 4.2 Employment structure of the Eindhoven region, 1994–96, number of people employed

Source: IVA, 1997.

As well as some well-known industrial firms, such as Philips Electronics and DAF, the region also hosts some fast growing high-tech companies such as ASMLithogaphy, Simac and Neways. The industry is very modern and knowledge-intensive. Compared to other important industrial regions in the Netherlands, the innovativeness of Eindhoven's industry is very high: 50 per cent of total Dutch R&D expenditure is laid out within the region (Kusters and Minne, 1992). With these figures, Eindhoven ranks among the top three of most knowledge-intensive regions in Europe, together with Paris in France and the Munich/Nürnberg region in Germany (NEI, 1996). The high knowledge-intensity of the region is generated by large firms such as Philips ad Nedcar, which spend enormous amounts of money on R&D. But regional knowledge institutes such as the Eindhoven University of Technology, institutes for higher education and private research firms also contribute to the innovativeness of industry.

NEI (1996) distinguishes a number of industrial 'knowledge-clusters' in the region of Eindhoven/Venlo: see Table 4.3. The most important clusters in the region are electronics, automotive and office equipment. The dominant position of knowledge-intensive industry in the region serves as an engine for other sectors, such as trade, business services and other services, and forms an important constituent of the mechatronics cluster, as will become clear in the next sections.

Table 4.3 Knowledge-intensive industrial clusters, Eindhoven/Venlo, people employed in different categories, 1996

	Producers	Suppliers	Knowledge infrastructure	Total
Electronics	6,000	9,200	4,000	19,200
Automotive/transports	4,100	10,500	4,000	18,600
Office equipment	5,300	6,200	4,100	15,700
Transport/distribution	4,100	1,000	3,900	9,400
Environmental technology	–	370	3,973	4,343
Medical technology	5,100	2,000	3,600	10,700

Source: NEI, 1996.

Table 4.4 Largest firms (in terms of employment) in the Eindhoven region

Firm	Industry	Employees in Eindhoven region	Total number of employees	Total turnover (Dfl, millions)
Philips Electronics	Electronics	21,600	262,500	69,000
DAF Trucks	Automotive	3,700	4,959	3,000
Rabobank	Financial	1,865	40,152	331,000
Stork	Manufacturing	1,150	20,500	4,900
ASMLithography	Manufacturing	1,100	1,600	1,300

Source: Eindhovens dagblad, 1997b.

3 Mechatronics: Context and Definitions

In this section, the concept of mechatronics is defined and elaborated. Next, a distinction is made between the technological and the organisational aspects of mechatronics.

3.1 Definition

The word mechatronics is a combination of mechanics and electronics. It refers to the combination of precision mechanical engineering, electronic control and systems-thinking in the design of products and manufacturing processes. Mechatronics is not a new technology; nor is it a new industrial structure. It is basically a methodology, a way of thinking and working during the design process, where the design team tries to overcome the barriers between the disciplines of mechanical engineering, electronics, information technology and physics. Thus, although literally, the word 'mechatronics' is derived from mechanics and electronics, other disciplines may also be involved. In the mechatronics approach, a distinction can be made between the technical aspect of mechatronics and the organisational aspect. The two perspectives are discussed below.

3.2 The Technology Perspective

From a technology point of view, mechatronics is about combining different technologies to reach a certain end, regardless of the organisational aspect (see Figure 4.3).

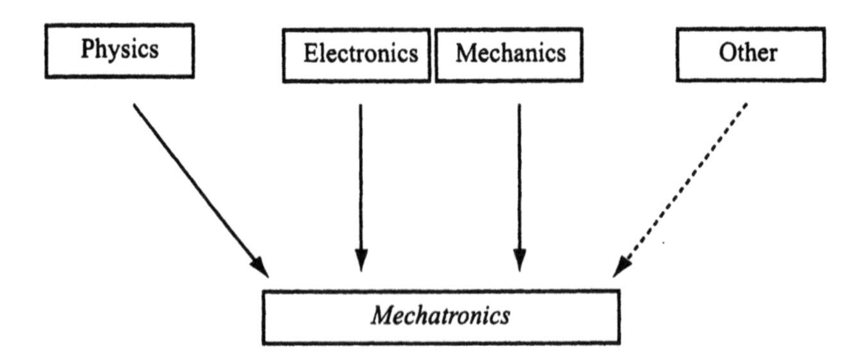

Figure 4.3 Mechatronics as combination of technical disciplines

In the technical sense, mechatronics refers not so much to the actual production of products as to the development and design of new products. Mechatronic design implies that during the design phase of a new product, specialists from several disciplines meet and choose the most efficient and effective mix of technologies to achieve it. The synergies arising from the mechatronic approach yield better technical results than the separate contributions of the different disciplines would. The advantage of mechatronics methodology is that a (technical) problem is approached from different angles, instead of from one technical discipline, to entail better technical solutions. The result of the mechatronic approach is a more efficient and effective design. Given the increased complexity of products, the mechatronic methodology is gaining popularity: increasingly, products incorporate many different technologies.

3.3 The Organisational Perspective

From an organisational perspective, the technical multidisciplinary approach may take different forms: firms can choose to develop and produce most things themselves, or specialise in one specific technology or part. Figure 4.4 shows a matrix with four possible organisation forms in mechatronic production and design.

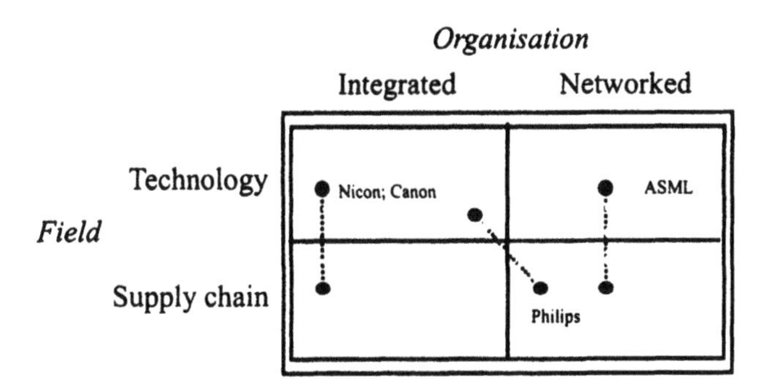

Figure 4.4 Mechatronics organisation matrix

The horizontal distinction is between the 'integral' approach and the networked approach. In the integrated approach, most of the activities are carried out within one, integrated firm. In the networked approach, cooperation structures and outsourcing relations with other firms characterise the way of working. Vertically, a distinction is made between technology and supply-chains. The technology aspect refers to the development of new technologies or new applications of existing technologies; the supply-chain (or production chain) represents the actual production process.

In the completely integrated approach – at technical and supply chain level – one large firm has all the technical specialisations 'in house' and is able to build teams of experts from within the organisation to design and develop a new product the mechatronic way. It is also able to produce all the raw material, semi-products and other supplies itself. This vertically integrated approach is adopted by some high-tech Japanese companies such as Canon and Nicon. The other extreme is the network approach. In this organisational model, different firms with one or more specialisation organise themselves in networks – with other firms, or with knowledge institutes – regarding both the technology and the supply chain. ASMLithography, a high-tech firm located in the Eindhoven region, is an example of such a firm.

The network-approach has gained popularity in the last decade, under the influence of fundamental developments. The most important is the increasing pace of technological progress. Technology progresses so rapidly that no single firm can master all technologies: at least some specialisation is unavoidable. This means that if a product is to be designed in which different technologies are incorporated, cooperation is a 'conditio sine qua non'. As a consequence, in the last decade firms have more and more concentrated on their 'core businesses': they do a few things they are good at and build on these core competencies. Large conglomerates have been broken down into series of independent profit centres or business units. Network structures are in many cases an appropriate answer to face challenges of the economy, in which international competition urges firms to cut costs, improve quality and enhance flexibility simultaneously.

Delimiting the scope The core part of this case study is the analysis of the mechatronics cluster in the Eindhoven region. However, because of the widespread application of mechatronic production and product design in many different sectors, it is not easy to delimit the scope of the cluster exactly. In modern industry, hardly any firms adopt only one technology. In most industrial products, different technologies are incorporated. This observation would suggest that any high-technology firm can theoretically be counted as member of the mechatronics family. It is illustrative in this respect that in a study on knowledge intensive clusters, the mechatronics cluster was depicted as comprising and surpassing other clusters in the region such as medical technology, automotive, office equipment and electronics (NEI, 1996). In that sense, mechatronics cannot be compared to other sectors but cuts through any high-tech industrial sector.

Given the scope of this research and to avoid rendering the sense of the mechatronics cluster meaningless, the concept was somehow restricted. As a starting point, a few important, large firms in the region which have explicitly adopted mechatronic methodology on a large scale were selected. Next, from our cluster perspective we took into consideration their main partners in production and product development, as well as their relation with the knowledge infrastructure in the region of Eindhoven.

4 Actors in the Mechatronics Cluster in Eindhoven

In this section, the actors in the mechatronics cluster and their activities are described, as well as the most important regional and local policy initiatives regarding the cluster. Section 5 is dedicated to the analysis of strategic relations amongst and between the actors.

4.1 Firms

Firms are the central actors of the mechatronics cluster. The number of firms involved in the mechatronics-cluster in the Eindhoven region is great. NEI (1996) estimates that there are 62 firms, employing a total of some 12,200 people. There are important differences between types of firms in the cluster. Some are very large, and operate on global markets, such as Philips; others are much smaller and have a more regional focus. To differentiate amongst firms, a distinction between types is depicted in Figure 4.5. Horizontally, a distinction is made between makers and thinkers; vertically between product- and capacity-oriented firms.

Within this division, the first category are the licence takers. They produce high volumes of products for large markets. This type of firms do not invest heavily in research and development. They rather use technology that has already been developed elsewhere, and apply it to their products. These firms are 'makers' more than 'thinkers'. This type of firms are relatively scarce in the Eindhoven region. Some 10 per cent of the firms in the mechatronics cluster are estimated to be active in this segment.

The second group, the licence providers, are the large and smaller industrial firms in the region that make substantial R&D efforts: they are 'thinkers'. Some 30 per cent of the mechatronic firms in the region are estimated to

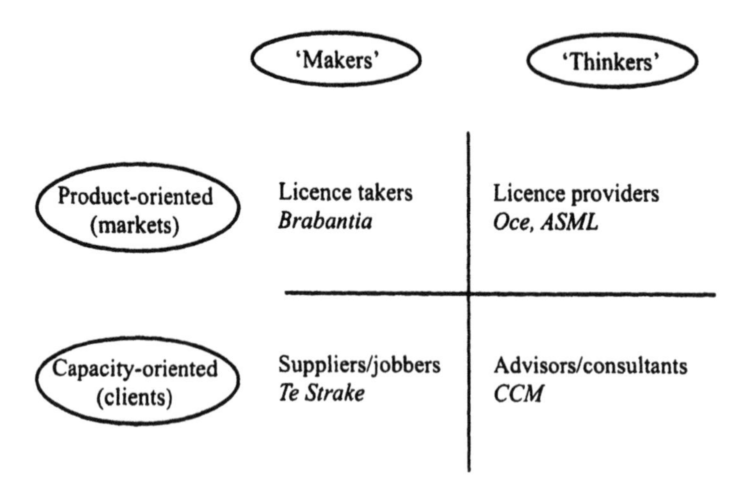

Figure 4.5 Firm typology

Source: Based on Van Gunsteren, 1996.

belong to this category. The most important are Philips (the electronics multinational), Océ (copiers producer), Stork (machine building), DAF (lorries) and ASML (wafer steppers). Table 4.5 shows some basic data on these firms.

Table 4.5 Large firms in the cluster

Firm	Products	Employees	Turnover (Dfl, billions)	R&D staff in Eindhoven region	R&D as % of sales
Philips	Electronics	267,200	80.0	1,700	7%
Océ	High tech printing and copying products and services	20,000	5.4	1,527	7%
ASML	Wafer steppers	2,400	1.8	681	11%
DAF	Lorries	5,800	n.k	460	–
Stork	Machines				
Nedcar	Cars	4,400	2.1	320	–

Source: Annual reports; Internet resources.

In most cases, the core activities of these firms are research and development and marketing. Other functions are outsourced. Research and development is executed partly within the borders of the firm, but to a large extent in cooperation with other firms. To that end, firms organise extensive R&D networks with technology partners. The large licence providers (often called original equipment manufacturers) in the Eindhoven region are fundamental 'engines' in the mechatronics cluster, as will become clear in section 5. Research, development and licence providing are not limited to these giants: many smaller firms put up substantial R&D efforts and participate in technology networks as well.

The third type of firm, the suppliers and jobbers, supply modules or parts to other firms, mostly to the licence providers. Within this category, there are large variations as to the technological level and sophistication of the firms. Some jobbers are very straightforward executors of standardised products: they get a set of specifications from the commissioning firms and produce a number of copies. Other supplying firms are much more sophisticated. They do not just produce a single part, but complex modules with a high technology content. Some of these firms are active in R&D as co-developers together with the commissioning firm. They are the so-called main suppliers, and form a vital constituent of the mechatronics cluster. The growth of this type of firm has been enormous, mainly through the inclination of large original equipment manufacturers to concentrate on core activities, outsource complete modules to reliable suppliers, and involve suppliers in product development. An example of a fast growing main supplier is Te Strake. In 1993, the turnover of this company was only Dfl 27 million. By 1997, it had risen to Dfl 66 million.

The fourth type of firm is the advisor/consultant in the region. The majority within this type are engineering bureaux and other types of technology advisors. In the Eindhoven region, there are around 100 of these bureaux. Another characteristic of this type of firm is that they work for a limited number of clients and not for large markets. A successful example within this category is CCM (Centre for Construction and Mechatronics). CCM carries out research and design activities for innovation projects of industrial firms. The approach is clearly mechatronic: technical specialists from different fields work together on integral solutions. In some cases, CCM also exploits its inventions commercially. The advisor/consultancy firms are 'supportive' to the mechatronic development and production firms described above. It is illustrative in this case that the majority of CCM's clients are located within the Eindhoven region.

4.2 The Public and Semi-public Knowledge Infrastructure

The second important type of actor in the mechatronics cluster are actors in the 'knowledge infrastructure': educational institutes and research institutes. Given the characteristics of the mechatronics cluster, the technical institutes are particularly important. In the region of Eindhoven, technical education is present at all levels, from secondary education to university level. In this section, the most important institutes will be described.

The Eindhoven University of Technology Probably the most important player in the educational field is Eindhoven's technical university. This university was founded in 1956. After the war, Dutch policy makers were convinced of the leading role of the industry in the reconstruction of the country: the number of technically skilled people at the highest level would have to be increased. The presence of Philips in the region, in particular the NatLab, was an important argument in favour of founding the new university in Eindhoven, although the Philips president at that time, Frits Philips, preferred Den Bosch (the capital of the province of Noord-Brabant) as a location for the university because of its presumed better 'culture' for students. Active lobbying and promotion of the local politicians in Eindhoven were decisive in getting the university to Eindhoven (Eindhovens Dagblad, 1996).

In the Netherlands as a whole, there are only three dedicated technical universities. The other two are located in Delft, in the western part of the Netherlands, and in Twente, in the east. The Eindhoven University of Technology employs 2,600 people, has seven faculties and offers 12 disciplines. Every year, some 1,000 students graduate. The university offers a complete range of education in many technical disciplines, and carries out a broad range of research, both fundamental and for the industry (see also section 6 for a further discussion on industry relations). At the Eindhoven University of Technology, a specific two-year postgraduate course is offered aimed at mechatronic designing. On this course, students with first degrees in mechanical engineering, electronics and physics are trained to develop skills to combine these disciplines and to think and work interdisciplinarily. An important institute attached to the Eindhoven University of Technology is the University Technology Institute (UTI), which is active specifically in finding multidisciplinary, mechatronic solutions for companies, and combining knowledge of other technology institutes.

Research: TNO-Industry A recent new element in the knowledge infrastructure is the research institute TNO-Industry. This institute carries out contract research for firms, very often in cooperation with technical universities. It aims at developing solutions in the field of product development, production technology and materials technology. The approach of TNO-Industry is mechatronic: technologies are combined to solve problems. It employs 400 people, most of them high level scientists. In 1997, the institute decided to move to Eindhoven, to the campus of the University of Technology. The reasons for this were twofold. Firstly, the majority of TNO's clients are situated in or very near the Eindhoven region; secondly, it is felt that TNO can realise synergy from its location so near the University of Technology, which may help to offer better solutions for its clients. Cooperation will take the form of knowledge exchange, staff exchange, and joint projects with the leading institutes Metals and Polyming. Furthermore, research cooperation is planned in the field of materials technology. The Stimulus subsidy scheme has contributed Dfl 15 million to attract TNO (see also section 4.3).

The Fontys Institute for Higher Education The Fontys Institute for Higher Education Eindhoven offers higher education for some 13,000 students. It is one of the largest institutes for higher education in the Netherlands. Fontys counts three faculties, and 25 disciplines. For the mechatronic cluster, the technical faculty is the most relevant, in particular the disciplines of mechanical engineering, electrical engineering, technical physics and higher informatics. Within the discipline of mechanical engineering, students can follow courses in mechatronics. In addition, multidisciplinary projects are carried out by students from the above mentioned disciplines.

School for Industrial Design and European Design Centre There are several other relevant elements in the knowledge infrastructure. The most important is the School for Industrial Design, and attached to this, the European Design Centre, which educate students in he designing of industrial products. As the importance of industrial design in new products is increasing, this institute is an asset for the mechatronics cluster in the Eindhoven region. Equally important are the lower-level technical education facilities such as institutes for vocational training, as they deliver new employees to the cluster.

4.3 Role of Government

Although not directly involved, local and regional governments are important enabling and stimulating players in the mechatronics cluster in Eindhoven. Two aspects of regional policy initiatives stand out. The first is the European stimulus programme for the Eindhoven region, and the second is the role of the semi-public Syntens organisation.

The Stimulus programme Stimulus is an economic stimulation scheme for the region of southeast Brabant. The stimulus programme was set up in the early 1990s to fight the economic crisis in the region stemming from the heavy restructuring of Philips, the largest employer in the region, and the bankruptcy of DAF Trucks. The European Commission contributed Dfl 144 million to the programme, up to 1997. In 1998 a new stimulus scheme was set up, Stimulus 2. This project is set to run from 1998 to 2001; the European Commission contributes Dfl 183 million. That same amount has to be put up by Dutch government agencies (Eindhovens Dagblad, 1997a).

One important role of Stimulus is to stimulate cooperation between different actors in the region (firms, institutes, educational institutes) by financing cooperation projects. Stimulus finances 25 per cent of an innovative project if two or more firms are involved; if one of these firms is of small or medium size, the contribution rises to 35 per cent. If a public research institute is also involved, the percentage increases to 45 per cent of the project. Within Stimulus, 123 projects have obtained grants. In total, more than 100 companies were involved. One of the spearheads of Stimulus 2 is the improvement of cooperation between developers, suppliers and outsourcers in the region. The technology-stimulation programme of the Ministry of Economic Affairs will contribute a further Dfl 20 million for this project (Stimulus, 1998a). Within the Stimulus 2 scheme, a total of Dfl 38 million in subsidies is available for the improvement and strengthening of the knowledge infrastructure in the region. Of this amount, Dfl 15 million was invested in attracting TNO-Industry (400 employees) to the region.

One well-known and innovative example of a Stimulus project is the KIC-project (Kennis Intensief Cluster; Knowledge-Intensive Cluster) of Océ, the copier producer. This project is aimed at stimulating the joint development of new products and technologies of Océ and its more than 100 regional suppliers, for the purpose of upgrading the skills level and the quality of co-production and co-development of the suppliers, and keeping more work in the region. Stimulus has played an important role in the success of the mechatronic cluster in Eindhoven. Very fundamentally, it has created an increased openness both between firms and between firms and knowledge institutes.

Syntens Syntens is the new name of an organisation created by a merger of the Innovation Centre and the Institute for Small and Medium-sized Enterprises (IMK; Instituut voor het Midden en Kleinbedrijf). The main role of this institute is to create conditions for economic development and innovation and to link firms to each other. Firms looking for a partner to develop a new product or technology can contact Syntens, which maintains relations with over 800 firms in the region, and has detailed databases about these firms at its disposal. Explicit attention is paid to the 'soft assets' of firms such as the management culture within the firms, for it is a well-known fact that in the establishment of successful strategic relations between firms, these assets are of great importance. In some cases, Syntens is the initiator of cooperation between firms. For example, DAF, a large lorry producer, was looking for a supplier of the front wheel suspension, a rather complicated part. Syntens linked a few firms in the region, which together were able to produce this part for DAF.

5 Analysis: Relations in the Cluster

This section is dedicated to the analysis of the strategic relations amongst the actors in the mechatronics cluster. The most important strategic linkages are depicted in Figure 4.6. The numbers in the figure correspond to the subsection numbers in which the relations are elaborated.

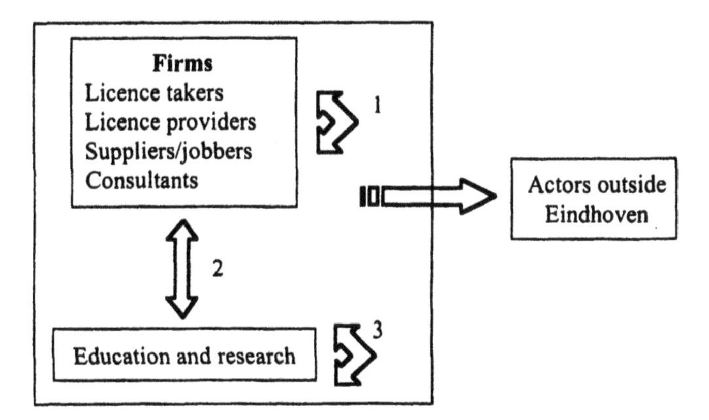

Figure 4.6 Relations and linkages in the mechatronics cluster in Eindhoven

5.1 Interfirm Linkages

As stated in section 3, mechatronic design and production can be organised within a large integrated firm (the integral approach) or in inter-organisational networks. In Eindhoven, the network approach is dominant, with Philips as an exception. In the Eindhoven region, the general picture is that strategic interfirm relations are strongly developed compared with other Dutch regions. The basis of these networks seems to be of a social nature: people from the different firms and organisations in the region know each other very well, as they frequently meet in many different manners, for instance at the many discussion platforms (such as the Mechatronics Platform and the Embedded Systems Platform), or informally at sports clubs or other unions. The density of clubs and associations in the region is one of the highest in the Netherlands. As a result, when a firm seeks cooperation, it is not difficult to find a partner in the region. Networking possibilities are further enhanced by the scale and diversity of the industrial base in the region: there is always a suitable partner. The intermediary activities of Syntens contribute to network formation by providing detailed information about possible cooperation partners in the region. The Stimulus Programme stimulates interfirm cooperation further, by providing financial incentives for innovative intra-regional cooperation projects.

When analysing strategic interfirm relations in the Eindhoven region more specifically, different aspects can be discerned. On a functional level, the most significant and intimate type of relationship is between original equipment manufacturers ('thinkers') and main suppliers in the region. As both types of firm increasingly engage in joint product development trajectories, the borders of the firms in some cases become blurred. Knowledge development and exchange is one of the characteristics of this relationship. In some cases, knowledge is further 'filtered down' to normal suppliers and jobbers. With a few exceptions – for instance, the main suppliers – the technological and organisational level of suppliers in the region is considered insufficient. The original equipment manufacturers regret this, as they would prefer to involve suppliers in development stages on a broader scale than they do now. An interesting development is that the high level main suppliers in the Eindhoven Region are trying to extend their market eastwards: they are seeking to establish strategic relations with large manufacturers in the nearby regions of Germany. This development could give the mechatronics cluster in Eindhoven new impetus.

In the spatial scale of networks, two dimensions can be discerned. There are technology networks on the one hand and supplier networks on the other, with a different geographical scope. In general, high-technology networks are organised on a world scale. In developing new technologies or applications, the large original equipment manufacturers (ASML, Philips, Océ) cooperate with technology partners which offer the best solutions, no matter where they are located. The suppliers' networks, however (i.e. the provision of modules or single parts), are much more concentrated and rooted in the Eindhoven

Case study: Global and Local Networks of ASM Lithography, Veldhoven

ASM Lithography (ASML) is a high-tech company that has grown very fast in the last few years. The firm is located in Veldhoven, near Eindhoven. With the Japanese giants Nicon and Canon, ASML is one of the leading firms in the world in the field of the development and production of wafer steppers (very complex and refined machines to build integrated circuits). ASML has a world-wide technology network: see the map below. Contacts with the overseas partners take place with modern communications means, such as videoconferencing. At the same time, not all problems can be solved in this way. Therefore, people from the technology partners are temporarily stationed within ASML. The suppliers network has a different geographical scope: the suppliers are much more concentrated in the Eindhoven region. ASML prefers doing business with firms in the immediate vicinity. According to ASML, the region of Eindhoven is well endowed with high-grade suppliers. However, the number of main suppliers (suppliers that participate in research and development phases of new products), is considered too small. ASML is searching for ways to upgrade the capacity and capability of current 'normal' suppliers, in order to be able to involve them in the development stage of new products.

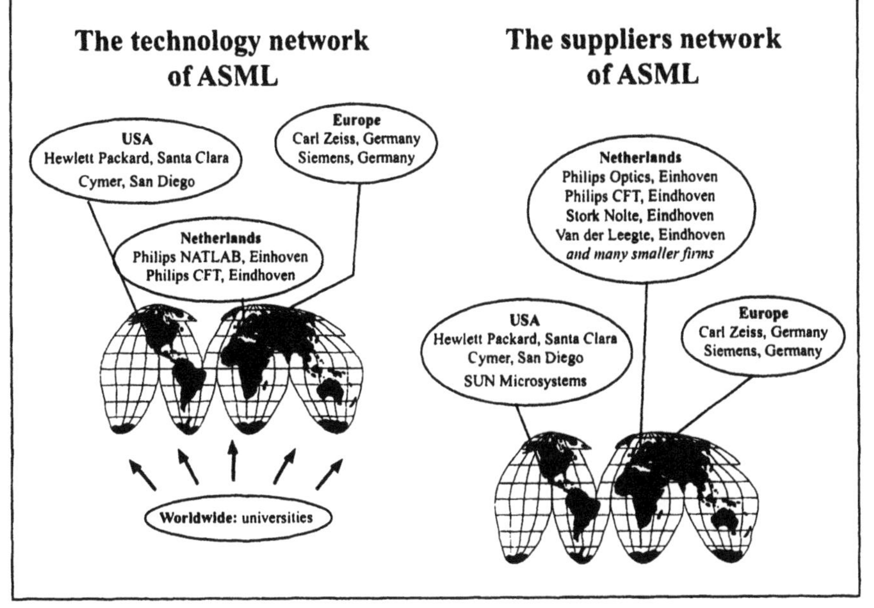

region. In the text box, the case of ASML is elaborated, but roughly the same holds for other firms such as Philips, Océ and DAF.

The role of Philips in the cluster In the Eindhoven region in general and the mechatronics cluster in particular, Philips plays an essential role, deserving special attention. Philips is the largest firm and largest technology player in the region. The most important parts of Philips are the Natlab (Natuurkundig Laboratorium; Physical Laboratory) and the Philips CFT. The Natlab, with an impressive 1,700 staff, provides for very high quality research. It is the most important research laboratory of the Philips organisation, and the largest private laboratory in the Netherlands. The other R&D centres of Philips are listed in Table 4.6.

Table 4.6 Philips research centres worldwide

Country	City	Number of employees
Netherlands	Eindhoven	1,700
Germany	Aachen	400
United Kingdom	London	270
USA	New York	250
France	Paris	190
Taiwan	Taipei	25

Source: http://www.research.philips.com/generalinfo/laboratories/.

In a recent study (NEI, 1996), the position of Philips in the mechatronics cluster of Eindhoven was elaborated. It was found that cooperative structures of Philips with other mechatronics firms in the region are relatively scarce, and on a small scale. The main cause is that the quality, capacity and scale of the firms in the region are often considered insufficient by the Philips organisation. Besides, the corporate culture of Philips is less inclined to external networking, but instead is traditionally biased towards keeping everything in its own hands.

Nevertheless, in a more indirect sense the role of Philips in the cluster is important. Spin-offs from the Natlab are considerable, in several respects. Knowledge from the Natlab spills over to other firms when people change jobs, which occurs very frequently. Also, the Natlab generates spin-off firms. This frequently occurs, when Natlab staff see market opportunities for a technology Philips does not want to invest in. An example of a spin-off firm from the Natlab is ODME, a company that produces CD and new media production equipment. Within a few years, this Veldhoven-based company has developed into a world leader as a producer of such equipment. It has a turnover of over Dfl 200 million, and employs 340 people, of which 260 are in the Eindhoven region (Eindhovens Dagblad, 1997b). In the third place, Natlab outsources some – though not many – activities to other firms in the region. These firms learn and benefit from the knowledge in the Natlab by cooperating in projects.

Officially and formally there is no technology and knowledge transfer from Philips to other firms in the region. In that respect it is illustrative that Philips does not participate in the cluster projects of Stimulus or in other regional initiatives. However, the situation is likely to change in due time. In 1997, the Philips organisation decided to move its headquarters from Eindhoven to Amsterdam. To compensate the region of Eindhoven for this heavy loss, Philips approved participation in a project named 'Philips in the region'. The aim of this project is to participate formally in knowledge and technology transfer projects. This project has the potential to increase the spin-off of the multinational electronics firm substantially

Additionally, at the current site of the Natlab, Philips plans to build a technology campus, where most Philips activities that are currently scattered through the city will be concentrated. Candidates for location at the campus are the semiconductor division, the CFT division and software departments. According to the current plan, the site will only be for Philips divisions, not for other firms. An exception may be made for ASMLithography, given its intensive contacts with Philips. Employment at the campus is to be doubled, from the current 3,500 to 7,500 in 2002. (Eindhovens Dagblad, 1998b). Many other firms are also eager to locate at the complex and benefit from Philips's proximity. The Chamber of Commerce is also very much in favour of an 'open' campus.

Missing elements in the mechatronics cluster Although the region of Eindhoven hosts a large number of varied firms in terms of size and function, some types of firms are under-represented. For one thing, there is a lack of firms that master optical technology at the highest level (although, there are some optical firms in the region and there is a programme at the university). This is a drawback, because optics is a vital element not only for some important firms such as ASML and Océ, but also for some smaller companies. For instance, ASML obtains its cutting-edge optical technology in Germany, from Carl Zeiss. Philips Optics, ASML's former supplier of optical technology, was considered to be insufficiently able to provide leading products. Océ buys its optical elements in Japan. A second insufficiently developed element in the mechatronics cluster is firms that are active in the field of embedded software and embedded systems. Within the Eindhoven region, the demand for these systems far exceeds the supply. To fill the gap, the University of Technology has initiated the Institute for Embedded Systems.

5.2 Relations and Linkages between Firms and Educational and Research Institutes

Linkages between the business community and the knowledge infrastructure are relatively well developed and certainly contribute to the success and dynamics of the mechatronics cluster. Four types of linkages strategic for the cluster can be discerned: the deployment of students in mechatronic firms; the execution of contract research for the industry; the collective setting up of platforms (scientific or technical); and the birth of new firms as spin-offs from education.

- In the first place, the firms in the mechatronics cluster highly value the presence in the region of technical education institutes on all levels as providers of a constant flow of new staff and a source of trainees. For instance, students of the postgraduate mechatronic designers course at the university carry out design projects in firms to practice their newly-acquired skills. Examples of design projects are the design of a thread break for an automatic loom at Te Strake (a main supplier), the implementation of a semi-active suspension for lorries at DAF Trucks, the design of a mushroom harvest machine (at CCM, an engineering bureau) and the design of a precision 3D-measurement machine. The Fontys Institute of Higher Education is also strongly linked to the regional industry, as it provides firms with new staff and trainees.
- The second type of linkage between firms and knowledge infrastructure consists of research activities carried on by the university (and Fontys) for mechatronics firms in the region. Contract research for the industry is important, as it benefits the knowledge institutes – they earn money and increase the relevancy of their research – and the firms. The Eindhoven University of Technology is active in that respect: attached to the university are many different technological institutes carrying out contract research for the industry. The transfer agency of the university, through which all contract research is channelled, has a turnover of some Dfl 30 million. The Fontys Institute also carries out contract research. For example, in

1997 Fontys developed a moulding technology, enabling the creation of aluminium moulds. This project was carried out by teachers of Fontys and a mould firm, and was sponsored by the Stimulus project. Fontys is particularly active in the support of small and medium enterprises (SMEs). For this purpose, the innovation workshop was established, in which students and teachers offer a broad range of services and solutions for SMEs. For instance, they advise SMEs about the use of new technologies, make prototypes of new products and sometimes help with the development of new products. The general aim of the initiative is to improve the use of the available knowledge and experience in Fontys, and to increase this knowledge by linking up with firms. The project also generates very useful and interesting thesis projects for students. An example of a mechatronic solution offered by the workshop is a project in which a hydraulic management system was developed for a firm, using laser technology, optical technology and a computer. The turnover of the innovation workshop amounts to some Dfl 425,000. During the last few years, this figure has been growing by some 6 per cent yearly (NV Rede, 1998).

- A third type of strategic interaction is the joint participation of education institutes and firms in platforms. A recent example is the foundation of the Institute for Embedded Systems by the university, with strong support from the business community. Embedded systems are 'intelligent' elements like hard- and software, which are increasingly incorporated – or 'embedded' – into all kinds of product. This initiative strengthens the mechatronics cluster, as the cluster seems to lack sufficient knowledge and capacity in this field. The institute will start modestly by coupling the activities of 12 professors of the Eindhoven University of Technology. However, the business community is deeply involved in the setting up of education and research programmes and plans to contribute financially in due time. Illustrative of the intimate links between the university and the business community is the fact that the initiator of this institute is a senior researcher at the Philips Natlab and professor at the university.
- Fourthly, education institutes generate dynamics in the cluster by providing a breeding ground for new firms. For instance, the Eindhoven University of Technology offers different facilities for newly-created firms. Such firms with a high technology intensity can settle on the university campus at reduced (subsidised) rents. The policy of the university is particularly directed at two kinds of firms: firms that can benefit most from the vicinity of the university because they use facilities and, secondly, firms with which the university closely cooperates or sees opportunities to cooperate with.

A recent initiative relevant for the cluster is the establishment of a so-called 'twinning centre' on the campus of the Eindhoven University of Technology. The aim of this centre is to offer space for young entrepreneurs in the field of information and communication technology (ICT), and 'twin' these firms with the expertise of senior business people to increase their chances of success. The initiative for this concept was taken by a senior manager of Philips and the Minister of Economic Affairs, with the aim of strengthening the position of Netherlands in ICT. It is a public-private partnership, as can be seen in Table 4.7, representing the financing of the centre. There are only two twinning centres in the Netherlands. One of the reasons why Eindhoven was successful in attracting the twinning centre was that the municipality succeeded in raising funds from different sources very rapidly.

Table 4.7 Contributors to the Eindhoven twinning centre

Actor	Contribution, in Dfl million
Ministry of Economic Affairs	2.0
Province	1.5
Municipality	1.5
Stimulus scheme	3
University of Technology	1.5
Firms	1.5

Source: Eindhovens Dagblad, 25 June 1998.

The large firms in the region (Rabobank, Philips, ASML, KPN) highly appreciate this centre, as they have an interest in strong ICT firms in their vicinity. The private sector contributes some Dfl 1.5 million to the centre. Philips invests Dfl 400,000. With the arrival of the twinning centre and TNO-Industry, the university campus will develop into a focal point for technological research and business.

5.3 Relations amongst Educational Institutes

The Eindhoven University of Technology and the Fontys Institute for Higher Education cooperate in different fields. On the highest level, the two institutes have delegates on the management team. On an operational scale, the Fontys Institute has a dedicated programme for students who were unable to be successful at the university. Furthermore, the institutes cooperate in some specialised technological fields to increase the transfer of research results into educational products. One example is integrated circuits. The knowledge about integrated circuits which is developed by the university is transferred to Fontys as soon as possible; this enables Fontys to provide the most up-to-date education and gives them a lead over other institutes.

A second initiative is called 'united brains, the network for solutions'. It is a network operated by the Eindhoven University of Technology, the Fonstys Institute and the design academy. It has three functions. The first is to carry out (or assist with) design projects for firms (for example, an autoCAD designer can electronically send his/her design to the Fontys Institute for comments or suggestions. The second function is the provision of knowledge and technology transfer (in many cases on-line). The third function is the provision of facilities such as virtual reality or supercomputer-computing. The project relies on an 'information superhighway' among the three institutes, with linkages to firms. The aim is to develop and present integral solutions for firms related to design, a kind of on-line help desk. The total costs of the project amount to Dfl 4 million. The three institutes have invested Dfl 2 million in total. Dfl 1.75 million are subsidies, of which Dfl 1.1 million from the Stimulus project. In the long run, the project will have to be self-supporting (Stimulus, 1998a).

6 Confrontation with the Framework

In this section, the cluster is confronted with the framework of reference developed to analyse the cluster in the context of the urban economy in general, and the organising capacity regarding the cluster and the integration of the cluster within the broader urban economic structure.

6.1 Spatial and Economic Context

The economic and spatial characteristics of the region of Eindhoven are, in general, favourable and conducive to the development of the mechatronics cluster in the region. To begin with, the economic structure of the region, with its long tradition and high bias towards high-tech and knowledge intensive industries, is very well developed. Furthermore, the region offers all kind of services and provision for the firms in the cluster.

A second element in the spatial-economic context of the mechatronics cluster is the 'cultware' in the region. There is a strong culture of cooperation: inter-organisational relationships are well developed, and often find their roots in social networks. The resulting high levels of mutual trust are conducive to

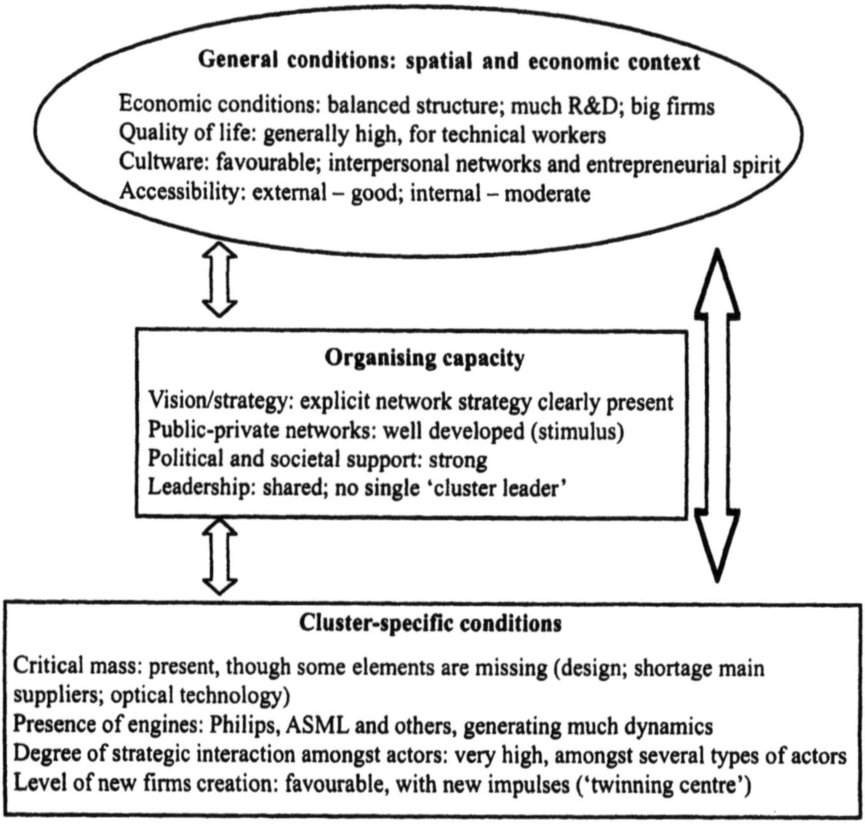

General conditions: spatial and economic context

Economic conditions: balanced structure; much R&D; big firms
Quality of life: generally high, for technical workers
Cultware: favourable; interpersonal networks and entrepreneurial spirit
Accessibility: external – good; internal – moderate

Organising capacity

Vision/strategy: explicit network strategy clearly present
Public-private networks: well developed (stimulus)
Political and societal support: strong
Leadership: shared; no single 'cluster leader'

Cluster-specific conditions

Critical mass: present, though some elements are missing (design; shortage main suppliers; optical technology)
Presence of engines: Philips, ASML and others, generating much dynamics
Degree of strategic interaction amongst actors: very high, amongst several types of actors
Level of new firms creation: favourable, with new impulses ('twinning centre')

Figure 4.7 Framework of reference

innovation, as trust is necessary to cooperate in innovation. In particular in mechatronics, where technologies are combined and specialists from different fields meet, a culture of trust and shared identity is helpful in breaking traditional barriers and achieving results.

The spatial characteristics in the Eindhoven region are favourable for the cluster. This holds for quality of life as well as for the accessibility of the region. Firstly, the location climate of the region on the whole is very favourable for mechatronic firms and their employees. By Dutch standards, the Eindhoven region offers ample housing space. Additionally, much natural beauty and open space surround the city of Eindhoven. Other quality-of-life aspects are less developed in the region. For instance, the city of Eindhoven lacks a metropolitan atmosphere. Cultural provisions are developed to some extent, but compared to large European cities their quantity and quality are not impressive. For technical employees, by far the most important category for the mechatronics cluster, the presence of nature and space are important factors, and seem to outweigh the disadvantages of a lack of metropolitan ambience. On average, mechatronics firms in the region have little difficulty in attracting staff relative to other Dutch regions. Even highly skilled foreign employees are generally pleased to live in the region. An important amenity in that respect is the presence of six golf courses, particularly appreciated by Japanese employees. The lack of metropolitan ambiance, though apparently not hampering the development of the cluster, implies that certain types of 'artistic related' economic activities such as multimedia firms and (graphical) design are under-represented in the region. In the long run, this could be a disadvantage for the mechatronics cluster, as design and styling are gaining importance in new products.

The international accessibility of Eindhoven is a very important issue for the mechatronics firms, which often operate on a European or even world scale. Firms are generally not unsatisfied in this respect, with the airports of Schiphol one hour's drive away, and also the international airports of Aachen (Germany) and Brussels (Belgium) nearby. Furthermore, there will be an easy connection with the future high-speed train link running from Amsterdam via Rotterdam to Brussels and Paris. The small airport of Eindhoven also fulfils a role in that respect. The former flight service of Philips has broken down. Several very small private airlines have emerged out of it, providing fast connections to Schiphol Airport, Zaventem and some other destinations.

The internal accessibility in the region falls short, however. The access roads into the city are not suited for the amount of traffic they have to carry. The road infrastructure of the city, a conglomerate of five former villages, is not adapted to current needs. The external accessibility of the region is good. The region is well provided with highways, and has an intercity rail station from which many destinations can be reached directly.

6.2 Cluster-specific Conditions

As has become clear from sections 4 and 5, the mechatronics cluster is well developed in several respects. In the first place, the cluster has some strong 'engines': large, internationally operating firms that act as a catalyst for other firms, both by outsourcing production but also by transferring state-of-the-art knowledge and technology to smaller and medium sized firms in the region. Second, strategic network relations within the region amongst firms and between firms and research/education are well established. The actors are mutually reinforcing, benefit from each other's presence. If the integration of the Philips organisation (in particular the large Natlab and CFT) with other firms in the cluster could be enhanced, the potential for further dynamic development is even more promising.

Secondly, the cluster has critical mass. The cluster is large, with many firms, a rich diversity (many different specialisms) and a broadly developed knowledge infrastructure. The advantage of this critical mass is that there is room for specialised services such as engineering bureaux, technical design agencies, consultancy firms, etc., which in turn are attractive to other firms and organisations considering relocation. Furthermore, this mass and diversity, in combination with the favourable 'cultware', results in an environment in which the newest (technical) knowledge in different fields is rapidly dispersed and adopted, thereby contributing to further dynamics of the cluster. The dynamics are further enhanced by the relatively high level of new firm creation, due to high levels of entrepreneurial sprit and stimulating public policies.

Despite the mass and diversity, some important elements are lacking. In terms of technology, the most important are state-of-the-art leaders in optical technology and embedded systems technology. Further, the region seems to lack main suppliers; firms that have sufficient skill and capacity to co-design and produce complex modules for the large original equipment manufacturers.

6.3 Organising Capacity

In the organising capacity related to the cluster, several aspects can be discerned: the presence of vision/strategy; the quality of public-private networks; the level of political and societal support for the development of the cluster; and leadership. Generally speaking, the organising capacity regarding mechatronics is well developed.

In the first place, the organising capacity of private firms in the regions is high. They are very eager to develop flexible networks and often prefer to work with partners in the region. In many cases, they take the lead in initiating of new activities, such as the setting up of specific platforms, while also involving the university and/or other organisations in the regional knowledge infrastructure. Second, there is an explicit and broadly shared vision about how business should develop in the region. Although this vision does not focus particularly on the development of the mechatronics cluster, the emphasis on networks creation and integration of business and research infrastructure is conducive to the mechatronics cluster. It is illustrative how much effort was made to attract TNO-Industry into the region: TNO was clearly regarded as a strengthening element in the cluster.

The Stimulus programme has been very helpful in regional network creation, as it stimulates firms to cooperate (with each other and with knowledge institutes) and provides financial incentives. The question remains whether networks will continue without these incentives in the future, but there are indications that many firms have discovered the advantages of networking and clustering.

Another important aspect is leadership in the development of the cluster. Although there is no single governmental body is integrally responsible for the mechatronics cluster in the region, cooperation between public bodies is generally smooth. The Stimulus project can again serve as an illustration. In the near future, the quality of organising capacity will be critical with regard to the upgrading of 'normal' suppliers to main suppliers. This is important not only for the large outsourcing manufacturing firms, which would benefit from the upgrading of their suppliers, but also for the region as a whole, because the presence of main suppliers is an significant location factor for other firms, being a catalyst for regional development. The question of who should take the lead in the upgrading process remains unanswered. The key will be to identify and implement business-oriented public-private solutions. Another issue requiring strong organising capacity is the integration of the Philips organisation into the region. As the opportunities that arise for this are substantial, a careful and prudent approach is needed. Philips has indicated a willingness to transfer more knowledge in the region, but under strong restrictions.

7 Conclusions and Perspectives

In the Eindhoven region, strong social ties and a high-tech tradition provide fertile ground for networked mechatronics. Large networked firms are the engines behind the cluster dynamics. They develop new technologies on a world-class level, with the world's top technology partners. State-of-the-art knowledge filters down from large firms to network partners in the Eindhoven region, resulting in a dynamic, learning region, with open windows to the world. Suppliers' networks exist on the regional scale, as firms generally prefer to outsource to firms in the physical and cultural vicinity. The knowledge infrastructure is very much involved in and part of the cluster, as it feeds the cluster with new talent, serves as breeding ground for new firms, and executes fundamental and contract research to the benefit of firms in the region, and is regarded as a very important location factor. The recent establishment of TNO-

Industry, the Eurandom Institute and the twinning centre in the region are further reinforcements of the mechatronics cluster.

The organising capacity in the region – vision, leadership, public private networks on several levels and political support – is one of the main 'silent engines' behind the prospering of the mechatronic cluster in Eindhoven. However, many organising skills will be needed to face a number of challenges. The first is to upgrade the technological and organisational level of suppliers, so that they can increasingly become involved in product development. Only a few (main) suppliers seem to reach a sufficient level. Constructive 'pre-competitive' cooperation is required between large outsourcing firms that would benefit most from excellent partners in their vicinity, but are separately too small to take up this task. The public knowledge infrastructure – university, higher education – could bring in its experience in knowledge transfer. Upgrading of the level of the suppliers is also in the interest of the municipal and regional authorities, as a means to tie the large firms more to the region and increase the local knowledge base. Therefore, some public-private form should be found to achieve the desired result.

A second challenge of organising capacity is to make more out of the presence of Philips in the region. The role of Philips can hardly be underestimated. It plays a fundamental 'tying role' in the cluster, generating spin-offs and knowledge transfer on several levels. The role of Philips could be exploited much more if the firm would participate more in regional networks. In this respect, as a first step it would be beneficial if Philips would allow other firms on the technology campus that it is currently establishing. Public incentives, such as the provision of additional infrastructure, could be helpful in this respect. Next, a more formal structure of knowledge and technology exchange between Philips and other actors in the Eindhoven region could further enhance dynamics, both of Philips and of its partners. The establishment of an independent intermediary structure, for instance led by TNO, could contribute to this.

References

Adang, A.J.V.M. and J.M.P. van Oorschot (1996), *Regio in bedrijf, Hoofdlijnen van industriële ontwikkeling en zakelijke dienstverlening in Zuidoost-Brabant.*
ASMLithography (1998), *Jaarverslag 1997.*
Eindhovens Dagblad (1996), *Hoop op herstel bij jubileum*, 23 April.
Eindhovens Dagblad (1997a), *Stimulus-2 definitief rond*, 30 May.
Eindhovens Dagblad (1997b), *ED bedrijven top-50*, 15 September.

Eindhovens Dagblad (1998a), *Mega computercentrum voor startende bedrijven*, 25 June.
Eindhovens Dagblad (1998b), *Natlab wil aansluiting op snelweg*, 23 July.
FPER Foundation for the Promotion of the Eindhoven Region (1997), *Eindhoven Region, Leading in Technology.*
Gunsteren, L.A. van (1996), *Management of Industrial R&D: A viewpoint from practice.*
IVA (1997), *Inzicht 1997: Sociaal economisch onderzoek regio Eindhoven.*
Kusters, A. and B. Minne (1992), *Technologie, marktstructuur en internationalisatie: de ontwikkeling van de industrie*, Den Haag, Centraal Planbureau.
NEI (1996), *Economische betekenis van industriële kennisclusters in de regio Eindhoven/Venlo*, NEI Rotterdam.
N.V. Rede (1998), *Jaarverslag NV Rede 1997.*
Polytechnisch Tijdschrift (1997), *ZuidoostBrabant snakt naar landelijke erkenning*, November.
Stimulus (1998a), *Nieuwsbrief Januari 1998.*
Stimulus (1998b), *Nieuwsbrief Maart 1998.*
Technische Universiteit Eindhoven (1998), *Jaarverslag 1997.*

Discussion Partners

Mr D.C. Boshuizen, TNO-Industry, Director.
Mr Bouman, The Stimulus Programme.
Mr Claessens, Fontys Hogeschool and Stimulus Programme.
Mr G. van de Kerkhof, Te Strake, General Manager, Business Unit Engineering and Production.
Mr J.W.M. Kummeling, CCM, Director.
Mr P. Nagel, Economic Affairs, municipality of Eindhoven.
Mr T. Schurgers, Syntens, Director.
Mr R. Vos, ASMLithography, Manager Materials Management.

Chapter Five

The Telecom Cluster in Helsinki

1 Introduction

This case study describes and analyses the telecom cluster in the Helsinki region. At first sight surprisingly, the peripheral city of Helsinki is one of the major concentrations of telecommunication and related activities in Europe, with Nokia, a well-known world leader in the production of mobile phones, as its most outstanding firm. Central to this case study is the description and analysis of the actors in the telecommunications cluster –firms, education, research, government – and their mutual relations, and the dynamics of the cluster. In addition, the cluster is put in the broader context of the economic, spatial and social structure of the Helsinki region as a whole.

This chapter is organised as follows. After an introduction in the regional-economic situation of Helsinki, in section 2, the main characteristics and trends of the telecommunications sector are discussed in section 3. Section 4 describes the principal players in the Helsinki cluster, as well as locational developments. Section 5, the pivot section, is dedicated to the analysis and judgement of the strategic interaction between the different actors. In section 6, the cluster is put in the social and economic context of the city. Very importantly, the organising capacity regarding the cluster is evaluated against that background. Section 7 concludes.

2 Helsinki: Profile of the Region and Economy

The city of Helsinki is located in the northeastern part of Europe. With a population of 532,000 it is by far the largest city in Finland. Three geographical entities can be discerned in the Helsinki area: the City of Helsinki, the metropolitan area of Helsinki (comprising Helsinki, Espoo and Vantaa), and finally the region of Helsinki, with eight more municipalities. In terms of financial and political power, the municipalities are the most important unit, as they have a high degree of autonomy and generous financial means. Local income taxes amount to 16 per cent of the total tax burden, a very high figure in European perspective. Some 62 per cent of the budget of the city of Helsinki comes from local taxes (Holstila, 1998). Dependency on the central state government is low: only 3 per cent of Helsinki's budget comes from national government funds. The metropolitan area – consisting of Helsinki, Espoo and Vantaa – has no government structure of its own: it is a platform for coordination of urban transport and waste management matters. Also on the regional level, there is no governmental body responsible for the region as a whole; nor are there regional elections. The regional council consists of members of the cities that constitute the region.

2.1 Population

Helsinki is the centre of the Helsinki region, which is a region with a population of well above 1.1 million people (see Table 5.1).

Table 5.1 Population in Helsinki and Helsinki region

City of Helsinki	539,000
Metropolitan area	920,000
Helsinki region	1,154,000

Source: The City of Helsinki Urban Facts, 1998a, p. 11.

Finland as a whole has only slightly more than five million inhabitants, which implies that more than one fifth of the Finnish population lives in the Helsinki region. The population of the Helsinki region is increasing. In the last few years, there has been a yearly migration surplus of some 10,000. For the years to come, the population is expected to grow further. From 1995–2015, growth is expected to be 24 per cent, which makes Helsinki the second growth region (in terms of population) in Europe, after Lisbon (City of Helsinki Information Management Centre, 1996).

2.2 Economic Development

The city of Helsinki was founded in the middle of the sixteenth century. Up to the nineteenth century however, the city remained insignificant. In that century the city began to develop more rapidly. By the turn of the century, it had grown into an important centre of services, commerce and administration. The industrialisation of the economy took off after World War I, and was accompanied by a rapid growth in commerce and services. After World War II, this process continued, although employment in the services sector grew much faster than in manufacturing. By the end of the 1980s, nearly half the jobs in the region were in finance, insurance, business services and public services.

During the last two decades the economic development of the Helsinki region has been characterised by large dynamics. Up to the late 1980s, the Helsinki region prospered. Income levels were rising steadily and the unemployment rate was very low, at only 3 per cent. In the early 1990s, the situation worsened dramatically, the main reason being the collapse of the Soviet Union, at that time a very important trading partner. Exports to the SU stagnated, while at the same time almost all of Europe was hit by a recession. In the period 1992/93, Finland's gross domestic product (GDP) dropped by 10 per cent and the unemployment rate rose to 18 per cent. From 1994 on, the economic situation has improved. Growth rates of GDP rank above the European average. GDP growth rates have been around 6 per cent per annum lately, regional growth in the Helsinki region even somewhat higher, and the unemployment rate shows a clear downward trend. For Finland as a whole, the unemployment rate is 10.2 per cent. The Helsinki region scores somewhat better, with 7.2 per cent. However, the fairly high number of long-term unemployed is a special problem in Helsinki. It has proved difficult to bring these people permanently back into active working life.

The severe economic crisis of the early 1990s did not bring only misery. It was an important stimulus for the actors in the region to join forces and combat the crisis; momentum was created to achieve fruitful cooperation. As a result, in 1994, the key actors in the region – administration on different levels and the business community – developed an integral and broadly shared vision, and drew up a regional strategy. Key elements of this strategy were to develop Helsinki as a Nordic region of interaction, with a focus on sustainable, knowledge-based development. Many plans were developed to give substance to this strategy. Examples are the setting up of science parks, the Art and Design city, and the Innopoli cluster west of Helsinki, but also many initiatives aimed at cooperation between business and research institutes were taken. The subject will be elaborated in section 5.

2.3 Economic Structure

The Helsinki region is the main economic centre of Finland, with an employment population of 514,000. The municipality of Helsinki alone counts more than 40,000 enterprises. In Table 5.2 the distribution of jobs amongst different sectors is depicted, for Helsinki, the metropolitan area, the region and Finland as a whole. The Helsinki region's economy is strongly dominated by services; some 80 per cent of employment falls into that category. Manufacturing is much less important, in particular compared to Finland's average. The high share of public services stems from the fact that almost all Finnish public bodies are located in Helsinki.

Table 5.2 Jobs by industry in Helsinki, metropolitan area, region and Finland, 1996, in % and absolute totals

	Helsinki	Metropolitan area	Helsinki region	Finland
Manufacturing	10.9	13.1	14.6	21.6
Construction	3.5	3.7	4.1	4.9
Trade, hotels, restaurants	17.7	19.8	19.5	14.5
Transport and communications	9.3	9.5	9.0	7.5
Finance, real estate and business services	20.2	19.0	17.7	11.1
Public and other services	36.7	32.7	32.4	31.1
Other, and unknown	1.7	2.2	2.6	9.3
Total	100%	100%	100%	100%
Total (absolute)	297,932	445,819	513,585	1,932,752

Source: City of Helsinki Urban Facts, 1998b, p. 45, adapted.

Despite its declining share in the economy, manufacturing is still important for Helsinki. Within the manufacturing sector, food products, paper and machinery are the most important branches. Industry is concentrated in the neighbouring cities of Helsinki, in particular Vantaa.

The Helsinki economy is relatively knowledge intensive. Research intensity (which can be measured in several ways) has increased in Finland throughout the 1990s. The region of Helsinki is by far the most important locus for R&D activity in Finland: over 50 per cent of Finnish R&D expenditure is spent there. Private involvement in R&D is high: the business sector accounts for some 62 per cent of R&D investments (City of Helsinki, 1998a). The knowledge intensity of the Helsinki economy can also be illustrated by the high educational level of its population. Some 50 per cent of Finland's academics live in the Helsinki region.

3 Telecommunications: Trends and Developments in Europe and Finland

In this section, some general developments in the telecommunications sector will be briefly discussed, followed by a short review of telecom developments in Finland. It serves as a background for the analysis of the telecommunications cluster in the Helsinki region, which is the main part of this case study.

3.1 General Trends in Telecommunications

The growth of telecommunications products and services has been tremendous in the last few years throughout the world and this growth is likely to continue. One of the fastest growing segments is mobile communication. The use of mobile phones worldwide grew from 135 million subscribers in January 1997 to 200 million at the end of that same year (Nokia, 1998).

The main causes of this growth are economic as well as technological, with an important interplay between economic and technical developments. A major economic factor has been the liberalisation of telecommunication (telecom) markets. In the recent past, in most European countries, telecommunication has been regarded as a state activity. Most telephone companies in Europe were (some still are) state monopolies. Recently however, there has been a strong wave of privatisation through the continent, caused by liberal policies and the generally shared vision that the provision of telecommunications services is not necessarily a state task, but can be left to private actors. European industry policy also strongly favours competition: telecom markets in all EU member states are to be privatised along more or less similar lines, in a several stages. In practice, however, the degree to which telecom markets have been liberalised differs widely among countries. The most liberal markets can be found in Scandinavia. More recently, Britain, the Netherlands and Germany have begun the process. The southern European countries lag somewhat behind. Privatisation and liberalisation have led to a wave of mergers, acquisitions and strategic alliances of new and formerly state-owned firms. A 'massive push' to create high bandwidth networks, capable of carrying voice, video and data is fuelling a surge in merger and acquisition activity during the last years. This holds not only for the telecom service providers, but also for the communications equipment industry (*Financial Times*, 1998).

In Figure 5.1, the most important global telecom alliances are depicted. National operators (often active in international consortiums) follow different

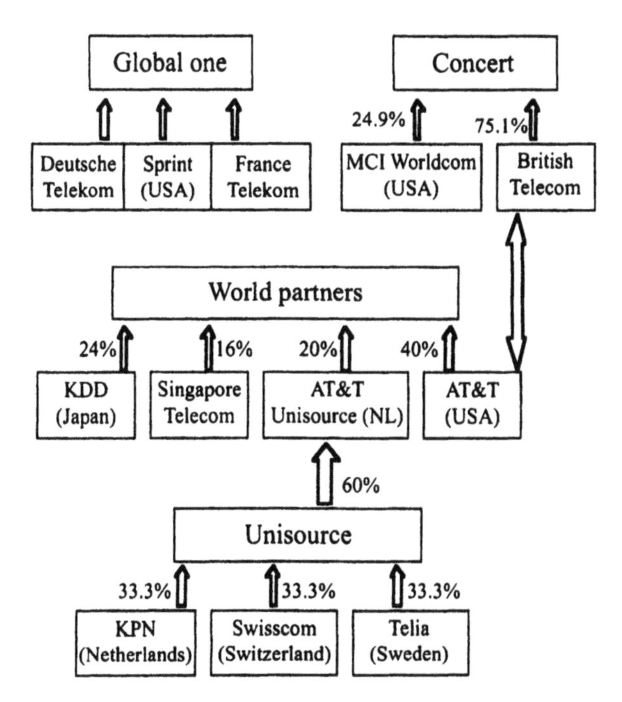

Figure 5.1 Major alliances in telecommunications

Source: NRC Handelsblad, 1998.

strategies. The largest (such as British Telecom) operate at a global scale, and invest heavily to become active in foreign markets. Smaller firms have more moderate intentions, and often engage as nondominant partners in international alliances with big players. The smallest telephone corporations often concentrate on niche markets (Finnet Group, 1996). A high degree of dynamics is generated by new players in the field. For example, in many countries energy companies have become involved in telecommunications (such as VIAG Interkom in Germany), or industrial conglomerates (Mannesman, Germany).

In general, liberalisation has led (and is still leading) to a much richer diversity in products and services offered. Competition has increased dramatically in the provision of telecom services. Private telecom companies have an incentive to provide the best services for their clients; they are more flexible and innovative than their state-owned counterparts. Moreover, it is generally felt that competition has increased the pace of innovations in telecom services. In many countries, the liberalisation has gone hand in hand with the introduction of new services, and prices have dropped considerably as well.

Simultaneously, other factors play a role in the upswing of the telecommunication sector. For instance, the increasing globalisation of the economy has increased the demand for (international) communication services considerably. But also the shift towards a network economy, in which interaction between people and firms is such an important determinant of success, can be said to give an impetus to the telecommunications sector.

Partly as a result of high demand and liberalisation and partly as an 'independent' process, technological developments have been very fast in the telecom sector, and create a high degree of dynamics. For instance, the rise of new interactive media such as the Internet and video-conferencing has created a new demand for data transmission. Another outstanding example is the relatively recent development of mobile communications. Until recently, mobile communication was very costly and complicated and only used for army, navy and yachting purposes. Rapid technological improvements and price cuts have made mobile communication available to almost everyone.

For the near future, the growth potential for telecommunication products and services is enormous. Firstly, application of (mobile) telecommunication will extend to many fields (for example health, teaching, working, car systems, and so on), which will boost demand for communications equipment and services. Secondly, technological developments continue to proceed at a rapid pace and will induce new demand. For instance, the third generation mobile communication standard is about to be introduced. With that technology, not only voice but also data and images can be transferred. Another development is the integration of mobile, fixed and data services. Traditionally, these services were offered separately, but increasingly they can be served by a single operator, in packages best suited to specific customer's needs (Nokia, 1998). Yet another important trend in telecom is the increasing awareness of the needs of the client. Much effort is put into increasing the user friendliness and the design of (consumer) equipment. The role of design in manufactured products alters the way manufacturers work, and requires that firms develop new capabilities. This point is elaborated in section 5.

3.2 Telecommunications in Finland

Finland is a very advanced country when it comes to the production and application of state-of-the-art telecommunication services and new media services. Finland has the highest penetration rate of mobile telephones in the world, exceeding 50 per cent in 1998, and Internet and new media services are also relatively widespread (see Table 5.3).

Table 5.3 Finland leading in mobile phones and internet penetration

Number of computers connected to the Internet per 1,000 inhabitants, 1998	Penetration of mobile phones, as % of population, 1997
1 Finland, 91	1 Finland, 41
2 Iceland, 89	2 Norway, 39
3 Norway, 73	3 Sweden, 33
4 Sweden, 46	4 Denmark, 30
5 Netherlands, 38	5 Italy, 19

Sources: Nokia, 1998; City of Helsinki Urban Facts, 1998b, p. 46, adapted.

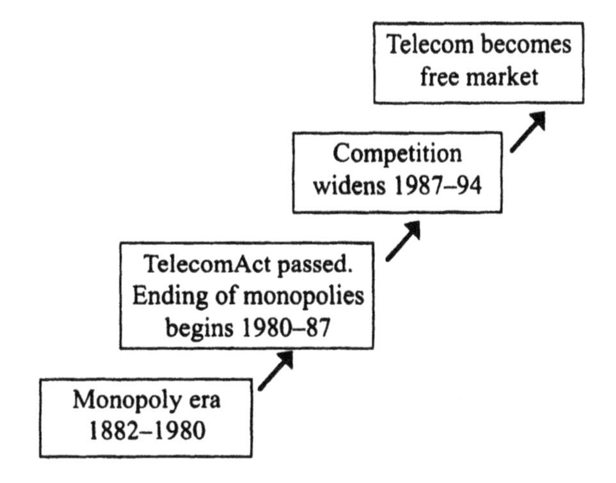

Figure 5.2 Liberalisation of Finnish telecom market, 1882–1998

Source: Finnet, 1996.

A basic reason for the high penetration rates of new communication devices, the low prices and the success of Finnish telecom service and equipment producers can be traced back to the early liberalisation of the Finnish *telecom* market. Finnish telephone operations have been based on private ownership from the start, in 1878. In 1882, however, the Senate, the authority at that time, wanted to control the telephone, which was considered to be a competitor of the telegraph (Finnet Group, 1996). The monopoly in telephone operations was abandoned in the early 1980s. In 1985, very early in the European context, competition was allowed in long-distance operations. After the Telecommunications Act of 1987 competition really took off, when telecom firms lined up against the General Directorate of Posts and Telegraphs (Finnet

Group, 1996). Competition gradually extended between 1987 and 1994. Now, almost the whole of the market in Finland is free.

The liberalised Finnish market was advantageous not only for telecom service providers, but also for equipment manufacturers. The demanding private clients (the telephone companies) had to compete to gain or retain their customers, and therefore needed innovative networks and equipment. In that way they gave local producers of telecom equipment – Nokia is the best example – an important incentive to innovate, and laid a basis for the current success of the cluster.

4 The Telecom Cluster in Helsinki: the Key Players

In the telecommunication cluster in the Helsinki region, several elements or 'actors' can be discerned. The most important are firms (different types), research institutes and intermediary organisations. Public policies regarding the cluster on different levels will be described as well. Given the important relations between telecommunication players (equipment manufacturers, telephone companies) and adjacent activities, this study not only takes pure telecommunications actors into account, but also the closely related sectors of new media, software development, media/art and design. The traditional media firms (publishers; radio and television; film) are left out; they are important users of telecommunications, but have fewer strategic linkages with actors in the telecom cluster.

4.1 Firms

Firms that are directly involved in the telecom cluster can be divided into several categories. The most important are the manufacturers of telecommunication equipment (such as mobile phones, switching equipment), and the providers of telecommunication services, such as telephone network operators. But also other types of firms play a role in the cluster, such as software development firms. Table 5.4 gives a very first impression of the size of the telecommunication cluster in the Helsinki Metropolitan Area.

Production of telecommunication equipment In the production of telecommunication equipment, one player is dominant in the Helsinki region: Nokia. It is estimated that Nokia constitutes half of the total electronics industry in Finland. The firm is famous for its development and production of mobile

Table 5.4 The telecommunications sector in the Helsinki metropolitan area, 1996

	Establishments	Employees	Turnover in FM (millions)	Share of Helsinki metropolitan area in Finland
Production of goods				
Office machinery	28	713	1,982	25.6%
Insulated wire and cable	9	183	27	9.0%
Radio, TV and communication equipment	89	5,629	6,736	25.7%
Production of services				
Telecommunications	100	5,898	6,218	44.2%
Computer and related activities	1,371	11,082	6,427	64.7%
Mechanical and process engineering design	272	1,839	1,131	24.0%

Source: City of Helsinki Urban Facts, 1998b, p. 30.

telephones. With a global market share of 21 per cent, it is a true world leader in the field. But it is also strong in telecommunication switching systems and other areas, such as car electronics and industrial electronics. With a share of 51 per cent in Nokia's total sales, the mobile phones division is the most important (Nokia, 1998).

Nokia has grown tremendously during the last few years. In the period from 1993 to 1997, net sales more than doubled, from FM 22,000 million to FM 52,000 million. This very successful firm has its headquarters in Espoo, located to the west of the city of Helsinki, and forms part of the Helsinki metropolitan area. Nokia employs 7,000 people in Helsinki, 22,000 in Finland as a whole, and 42,000 worldwide.

Nokia is very keen on the development of new products and tries to shorten time-to-market, as lifecycles of products shorten. It spends some 9 per cent of its net sales on research and development activities. One-third of its employees are active in R&D. The main part of Nokia's research is executed in the Nokia Research Centre, which is located at seven locations: two in Finland – Helsinki and Tampere – and five abroad (in Boston, Dallas, Bochum, Tokyo and Budapest). Furthermore, Nokia cooperates with 36 R&D centres all over the world, in 11 countries. Notwithstanding this international presence, the research activities are still concentrated in Finland. Fifty-seven per cent of the R&D jobs are in Finland, divided between four cities: Helsinki (the main centre), Tampere, Saloo and Oulu. The actual production of equipment is executed at other locations. In research, Nokia cooperates to a high extent with technical universities at the several R&D sites. See also section 5.

Although Nokia is the dominant player in telecommunication equipment manufacturing, there are very many smaller firms in the Helsinki region active in the field. Many of them have some links to Nokia, as suppliers of parts, components, software or other services. For a further discussion of the relation of Nokia with smaller firms in the region, see the next section. Foreign equipment manufacturers also have establishments in Helsinki. An example is Ericsson, the Swedish competitor of Nokia, which employs over 1,000 people in the Helsinki region.

Producers of telecom services The second important category of players in the cluster are the providers of telecom services, such as telephone companies of different kind and size, Internet providers and others. The major players in this category in terms of employment and turnover are the large telephone companies, of which several have their headquarters in Helsinki. Most of them only offer not only traditional telecom services (such as fixed telephone), but have also extended their product range with Internet services, mobile services and dedicated services for firms. The largest among them are Sonera and Finnet Group. Sonera is a former public telecom firm (its name used to be Telecom Finland), but was partly privatised recently. It employs some 8,000 people, and has its headquarters in Helsinki. The Finnet Group is a consortium formed by the 46 independent regional telephone companies. The core area of the group consists of traditional local telecommunication services (Finnet Group, 1996). One of the most prominent members of the consortium is Helsinki Telephone Corporation. This is the largest privately-owned provider of telecommunication services in Finland, with a turnover of FM 2,565 million and 3,814 employees in 1997 (HTC, 1998). Services provided include local, long-distance, international and mobile calls, data transmission and tailored planning and implementation of business telecommunications solutions. The firm is also active abroad, both in participation in telecom companies and in consulting new and existing telecom service providers in markets that are in the process of liberalisation.

The large telecommunication service providers are very important in the telecom clusters. Not only do they provide innovative services to firms and inhabitants in the city, they are also an important client of the producers of telecom equipment, such as Nokia and smaller firms. This point will be elaborated in the next section. Apart from these large operators, there are small Internet service providers and niche-players in the telecom market.

Others Other relevant firms in the telecom cluster are software companies and new media firms. Software-development companies are important, because telecommunications products increasingly embody software, or 'intelligence'. (In Table 5.4 they are included in the category of 'computer and related activities'.) There are very many of these firms in Helsinki. Often, their size is small. Secondly, new media companies play an important role in the cluster, both as demanders of innovative telecommunications services and as developers of new (interactive) products and services. It is estimated that the Helsinki region hosts more than 200 of these firms (70 per cent of the Finnish total!). Their number is growing tremendously. The average rate of growth in turnover within this sector is also impressive: about 375 per cent in 1995–96 and 100 per cent in 1996–97 (Media Studio, 1997a). Customers of these firms are buying new media services mainly for the purpose of corporate communication, internal communication and advertising (Media Studio, 1997b).

An example of a new media firm is Interaktiiven Satama (Interactive Harbour), which – among other activities – develops business solutions for the Internet, for many different firms. The Internet revolution offers so many possibilities and raises so many questions that firms like Interactive Harbour are very busy and grow very fast.

4.2 Education and Research

The telecommunication cluster is characterised by enormous dynamics, both in markets and in technology development. This is the reason why the knowledge infrastructure (education and research institutes) is of fundamental importance for the telecom cluster in the Helsinki region. In general, the region is well endowed with research and education facilities; see Table 5.5, in which the university level schools in the region are depicted, with their student numbers. For the telecom cluster, a few universities are of particular value. The first is the Helsinki University of Technology. This university is located in Espoo, very near the Otaniemi area, which is the high-tech location of the Helsinki region (see the next section). This is the main 'supplier' of technical staff for firms in the cluster, and also executes much research in the field of telecommunication products and services. Besides regular university education, it comprises a graduate school in electronics, telecommunication and automation. The Graduate School is a postgraduate programme offered jointly by the Helsinki University of Technology (HUT), the Tampere University of Technology (TUT), the University of Oulu (UO), and the University of Turku (UT). In all, the 29 participating departments and laboratories have 49 professors or associate professors, over 50 postdoctoral instructors and several lecturers (adjunct professors) from other universities. Part-time professors from industry and foreign visiting professors also participate in the activities of the school.

Table 5.5 University level education and number of students in the Helsinki metropolitan area, 1997

Institute	Number of students
University of Helsinki	33,419
Helsinki University of Technology	12,997
Helsinki School of Economics and Business Administration	3,789
Swedish School of Economics and Business Administration	2,170
University of Industrial Arts	1,455
Sibelius Academy (music)	1,421
Theatre Academy	332
Academy of Fine Arts	261
Total	55,844

Source: Ministry of Education, 1998.

Other important institutes for the cluster are the University of Helsinki, located in Helsinki's inner city, offering a very broad range of research and education activities. The University of Art and Design, though not very big, is also important. Designing skills are growing in importance, as the design of (telecom and new media) products becomes a determinant factor in commercial success. This holds particularly for consumer products, such as mobile telephones. The University of Art and Design has an Institute of Art and Communication. An interesting development is the Media Studio initiative. This is a pilot programme (running from 1996 to 1998) offered by the Art and Design Academy and co-funded by the Ministry of Education and the European Social Fund. It focuses on research into developments in new media and communications (together with the Helsinki School of Economics), and on training programmes in new-media production and design. The annual number of students – both professionals and nonprofessionals – exceeds 400. After 1998 the programme will have to be self-supporting.

Apart from universities, there are also polytechnic institutes, which are an important source of knowledge and staff: the Helsinki Polytechnic, the Haaga Polytechnic, the Helsinki Business Polytechnic, the Arcada Polytechnic (Swedish-speaking) and the Espoo-Vantaa Institutes of Technology.

4.3 Intermediary and Financial Institutions in the Cluster

There are several other actors which play a role in the cluster. In the first place, there are several intermediary organisations. Two examples are the Chamber of Commerce of Helsinki and the Culminatum prganisation respectively. The Helsinki's Chamber of Commerce is the largest of the Nordic countries, with over 4,000 members. Its main task is to promote the interests of the business community in the area, but it is increasingly broadening its range of activities towards offering training facilities for firms and giving all kinds of information and advice. One of its key projects is aimed at helping members to make the shift to the information society. The Culminatum organisation is another important intermediary structure. Culminatum Ltd Oy was established in 1995 to carry out the Centre of Expertise Programme – a national scheme aimed at stimulating knowledge and technology transfer between research institutes and firms – in the Helsinki region. Culminatum will accelerate the technology transfer process from ideas through research to saleable products by coordinating the collaboration between companies and scientists, and by undertaking a part of the financing for the projects. Many organisations –universities, municipalities, firms – are involved. With a budget of FM 10 million and only 10 employees, the means of this network organisation are very restricted. Another initiative is the Spinno programme, a training programme for entrepreneurs to turn business ideas to enterprises. It focuses on knowledge based enterprises. Several hundreds of people have attended the courses since the start of the programme in 1990.

Finally, financial organisations play a role in the cluster. Apart from the commercial banks, there are generic and specific public funds aimed at the stimulation of the information society. One example is the Finnish National Fund for Research and Development (SITRA). It provides venture capital – which is insufficiently supplied by the private banks – for promising, innovative firms, and also helps firms to commercialise new ideas and inventions. Being a national fund, it has no specific policy for the Helsinki region, but the largest part of its means go there. Another investment fund that is important for the dynamics of the cluster is named TEKES. This fund (from the Ministry of Trade), is specifically aimed at financing common projects of the industry and universities. Finally, linked to the Spinno programme, there is a venture capital investment fund named Spinno-Seed-Oy. It is owned by the cities of Helsinki and Espoo, Innopli Ltd (see next section), Sitra, and an SME (small and medium-sized enterprises) financing corporation. The fund is managed by Culminatum.

4.4 Locational Developments

Firms and institutes that are part of the telecommunications cluster in the Helsinki region can be found throughout the region, but there are two sites with a concentration of activities. The first is the area of Arabianranta, an emerging location in the city of Helsinki for advanced activities in the fields of new media, art and design; the second is the large and already well-established Otaniemi park in Espoo, a major site for innovative high-tech firms in the fields of information and communication technology.

Arabianranta/Art and Design City A very important locus of the cluster is the Arabiaranta site in Helsinki, located at 6 km from the centre of Helsinki. This large site (about 85 hectares) is intended to become a so-called 'Art and Design City' within the city of Helsinki. The Arabia district in Helsinki has been one of the centres of Finnish design for over 100 years. A few years ago the City of Helsinki and the other owners of the Arabia district inaugurated a large development scheme called the Art and Design City Helsinki Project. The scheme's aim is to turn Arabianranta into one of the leading art and design areas in Europe, where the functions of living, working, studying and recreation are mixed. The project is planned to be completed by 2014. The area already hosts 4,000 residents and 3,000 jobs; these numbers are intended to increase to 8,000 residents and 12,000 jobs.

An integral development strategy has been adopted, in which there is space for such complementary activities as research, business, education and housing and shopping. At its borders, it is surrounded by green leisure zones. At its core, it comprises an old pottery that is re-used as a 'design factory' where firms active in the field of new media and design are located. Already, 50 to 60 firms are present on the site, employing some 300 people. The number is growing, as the site is very popular. Rents are half those at the Innopoli site. Many of the small firms work for larger firms in the region, such as Ericsson, Nokia, Siemens and Sonera.

The University of Art and Design has been located on the site since 1985. It is one of the key elements of the project, as it offers research facilities, attracts many young people and acts as a breeding ground for young talents and new activities. The university is known to be business-oriented, and develops education and research programmes that comply with the needs of the business community. In due time, an audiovisual complex will be built near the university (FM 170 million). As an integral part of ADC new high quality housing and shopping facilities are being constructed. A concert hall with a capacity of 300 places is planned.

The area will also be used for housing. To create a pleasant residential environment, the surrounding landscape will be preserved: the Vantaanjoki river delta will be restored to its natural state, allowing recreational fishing and boating. A museum centre will be established as well. The electronic infrastructure of the art and design city will be very sophisticated. A region-wide broadband fibre network will be implemented, with open access for all residents, schools and businesses. The intranet offers opportunities for teleworking, -shopping and -banking, -health and -library services. Nevertheless, it is clearly recognised that despite all the potential of electronic media, face-to-face contacts remain fundamental in creative activities. The site is also attractive to large firms such as the Sonera Telephone Corporation and Nokia. These firms could benefit from the superior quality of the electronic infrastructure by using the area as a test-site for new products, or as a showcase to display what is possible in the future. The ADC is one of the 13 urban programmes that run in Finland.

The ADC is a clear public-private project. The main partners (and funders) are the Ministry of Trade and Industry, the City of Helsinki, Hackman Ltd (a large producer of consumer and design goods), Pop and Jazz Conservatory, and some other private firms.

Otaniemi Technology Park and Innopoli The second major site for the telecom cluster is the Otaniemi Technology Park, located in Espoo, a municipality west of Helsinki (but part of the Helsinki Metropolitan Area). This Science Park is at some 20 minutes driving from the centre of Helsinki and the Helsinki international airport. In the science park, about 5,000 researchers are employed. The Helsinki University of Technology, Geological Survey of Finland, the Technical Research Centre of Finland, the Finnish Pulp and Paper Research Institute and many more institutes of advanced and applied research are located here. There are nearly 200 companies in Otaniemi Science Park altogether.

The main building of the Otaniemi Science Park is Innopoli, which is one of Europe's most extensive commercial science park centres. In Innopoli the interplay and cooperation of science, product development and business activities takes place. It is an international marketplace for Finnish ability which also offers good connections to Scandinavia and Russia. The aim of Innopoli is to speed up the commercialisation and internationalisation of the results of investments in Finnish research and product development. Innopoli is a business site for more than 100 companies. A tenant unit or a company can find all the services it needs under one roof, including business services such as reception, telephone answering, post office, restaurant, auditoria, meeting rooms, a travel agency, data connection services, auditing, lawyers and marketing, as well as leisure services such as a gym, a sauna, a pub and a barber.

Innopoli's subsidiary, Otaniemi Technology Park, concentrates on incubation activities – the creation of companies founded on the latest technological skills. The Spinno activities executed here generate 20–30 high-technology companies annually, of which most continue their business activities within Innopoli. Otaniemi Technology Park concentrates on incubation activities – the creation of companies founded on the latest technological skills.

There is some degree of competition between Arabianranta/art and design city in Helsinki and the Innopoli site in Helsinki. For instance, Apple recently relocated from the Innopoli building to the Arabianranta. Generally however, the two are complementary, with Innopoli focusing on high-technology development and Arabiaranta on design-like activities (context).

4.5 Public Policies Concerning the Cluster

Public policies concerning the telecommunications cluster in the Helsinki region can be discerned on several levels: the municipal, the state and the European levels.

On the municipal level, Helsinki has no specific policies dedicated to the stimulation of the telecommunications cluster. The main reason is that the majority of actors in the cluster are located in the neighbouring city of Espoo. In general, the city of Helsinki considers the creation of conditions – a good living environment, good accessibility, fiscal climate and so on – its main task, and is not involved in active industry policy. Still, there are local policies which have an indirect impact on the cluster. An example is the participation of Helsinki in the Ariabanranta project, as described in section 4.4, and in the technology transfer activities of Culminatum. Similarly, the city of Espoo has contributed to the establishment of the Otaniemi park and Innopoli.

There are no policies on the metropolitan level either: Helsinki, Espoo and Vantaa only cooperate formally in the fields of transport and waste management. On the national level, there are policies in Finland favouring the cluster. One is the continuing liberalisation of telecom markets. Moreover, an important national programme – the Centre of Expertise programme – was launched to stimulate technology transfer from universities to SMEs. The Culminatum organisation was set up to carry out this programme in the Helsinki region. The Ministry of Education has increased the budget of higher education in the fields of telematics and information technology. Recently, the Ministry launched a programme aimed at coping with the threat of staff shortage in the near future in the field of IT. The programme is aimed at increasing the number of students in information technology and at re-educating (unemployed) people from adjacent specialities. The Finnish government invests FM 3 billion in this programme, but the private sector is also deeply involved, by providing equipment, laboratories, and advice. On the European level, finally, Finnish telecom firms benefit from many European technology programmes, in particular the 4th Framework EU programmes.

5 Interaction and Dynamics in the Cluster

In a dynamic and fast changing environment such as the telecommunication industry, rapid and frequent interaction and cooperation between the different actors in the cluster is of great value: a well functioning cluster enhances flexibility, permits a better use of the available knowledge, and boosts creativity in the region. Well-functioning strategic networks are likely to benefit all the parties, as they may generate a competitive advantage for the region as a whole. It is in that light that this section describes and analyses the strategic interactions between the different actors in the cluster. In Figure 5.1 the relations that will be described are schematically represented. The subsection numbers correspond to the numbers in the figure.

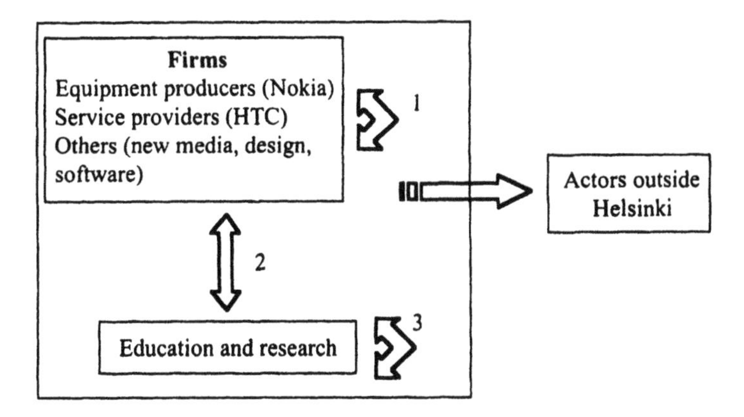

Figure 5.3 Relations in the cluster

5.1 Interfirm Relations

In the telecom cluster in Helsinki, interfirm cooperation of several kinds occurs very frequently. The limited scale of the Finnish market and the traditionally non-hierarchic, informal way of working in Finland favours strategic networking. Generally, people know and trust each other, and common interests are easily recognised. Examples of identified common interests are the high quality of technology and services, the need to create added value in telecom, a visible international presence and a sufficient amount of sophisticated know-how in the telecom cluster.

As it is impossible to describe and analyse all interfirm networks in the telecom cluster in Helsinki, we will focus on a few types of interfirm linkages that are of particular importance for the dynamics of the cluster: links between telecommunications manufacturers and services providers on the one hand, and links between large firms and SMEs on the other. Particular attention is paid to the position of Nokia as major firm in the cluster vis-à-vis other firms.

- *Linkages between manufacturers and service providers*. There are strategic contacts between telecommunications manufacturers on the one hand – the most important being Nokia – and operators on the others. Contacts between those two have been close for a long time. In Helsinki, the telephone companies' readiness to use innovative technology – inspired by the liberalised market in Finland – gave the industry the opportunity to develop new products and solutions. In other words: the strong and demanding home market created competitive advantage for the manufacturers (Finnet Group, 1996). In research and development, Nokia closely cooperates with telephone service providers Finnet Group and Sonera, as main clients of the Nokia systems.
- *Linkages between large firms and SMEs*. The large firms play a fundamental role in the cluster, not just because of their size, but also because they are tied into networks with many smaller firms. In particular, small firms in the cluster in the fields of new media and Internet have large firms in the region – Nokia, Sonera, Ericsson – as their main clients. In that respect, Nokia, as the biggest firm, deserves extra attention. Nokia has many relations with smaller firms in the region and generates many indirect spin-offs. In the context of EU technology programmes, it works together with many partners, not only large firms (such as Alcatel, Sonera, Ericsson) but also with high-tech SMEs, some of which are located in the technology park nearby. In that connection, it is important to note that in other Finnish

cities – notably Tampere and Oulu – Nokia subsidiaries are much more spatially and functionally integrated in the local science parks. The main reason is that, unlike in Helsinki, the science parks and the Nokia subsidiaries were established simultaneously at the same spot, and have 'grown up together' in these two cities, which has resulted in high levels of mutual trust and strategic cooperation. Although participating in many networks, Nokia does not regard itself as an *initiator* of networks with SMEs. In this respect, the firm has a relatively conservative attitude.

Recently, Nokia has begun to realise the importance of small technology firms. It has created a division that is to occupy itself with creating spin-offs of new companies. This division is active in three fields: developing new business units in strategic technology areas; financing new, promising firms; and generating new ideas for new businesses to start up. The general idea behind this is that Nokia has an interest in new firms that are active in fields relevant for Nokia, but which do not necessarily have to be incorporated in the Nokia firm. This development is promising for the future dynamics in the telecom cluster.

At the art and design city, with its concentration of small firms active in new media, an active policy is pursued which is aimed at linking the small firms to big firms (Nokia, Sonera, Ericsson) in the region. One of the objectives of this policy is to let the SMEs benefit from the international (client's) networks of the large firms, so that they can grow more rapidly.

5.2 Cooperation between Firms and Research/Education

As outlined in section 4, the principal players in the field of research and education regarding the telecom cluster are the Helsinki University of Technology (HUT), located in Espoo, and the University of Art and design. The role of polytechnics is substantial as well. In general, linkages between business and education are numerous. They certainly contribute to the success and dynamics of the cluster. Three types of linkages strategic for the cluster can be discerned: the role of firms in public education, the execution of contract research for the industry and the birth of new firms as spin-offs from the public knowledge infrastructure.

Firms and public education Private firms in the telecom cluster are very much concerned with higher education, in the first place because in the long run, the quality of graduates determines the future success of the firms. Secondly, the telecom cluster is growing so fast that there are labour shortages: well-educated technicians are very difficult to find. The two factors explain why the private sector is willing to co-finance a major state programme, worth FM 3 billion, aimed at increasing the number and quality of students of information technology and telecommunications. Apart from this financial contribution of the private sector to the development of human capital in the cluster, there are several types of linkage between university education and firms.

Firstly, in Finland, science students traditionally write their theses on the basis of a practical research project in a firm. The advantage is that students are better prepared for their future jobs and firms can benefit from their research and have some pre-selection of staff. At Nokia alone, there are 1,000 projects – organised and supervised by the research department of Nokia and the university jointly – in which students work on their Master's thesis. Furthermore, Nokia employs and guides students working on a their PhDs. In addition, PhD projects for fundamental research are financed by the Nokia Foundation. Although Nokia is dominant in the employment of students is outspoken – Nokia hires 50 per cent of the students of the Helsinki University of Technology – other firms in the region apply the same type of procedures to hire students.

In the non-science academic disciplines, thesis writing within a firm is much less common. Most students from the Art and Design Academy and the Helsinki University do not write their thesis on the basis of a business-related project. Nevertheless, these institutes do have relations with the local industry. For instance, Nokia has worked incidentally with psychology students from the University of Helsinki on small projects related to building brands and developing appropriate user interfaces for communication devices, and with students for the Art and Design academy in designing gifts. In the design of consumer products, such as mobile phones, Nokia does not work with the Art and Design Academy but with a top-level USA-based design firm.

In some cases, university staff are hired for in-company training in large firms – among which Nokia again figures. Alternatively, 'captains of industry' are frequently invited to lecture at the university. Nokia employees are also active in the composition of education programmes; Nokia staff have seats on several education committees of the Helsinki University of Technology. At the University of Art and Design, education programmes are also increasingly developed in cooperation with firms in the design sector.

Cooperation in research In the field of research, strategic links between universities and firms are frequent as well: in particular at the Helsinki

University of Technology, much contract research is executed for regional industry; the other universities are generally less business-oriented. In particular, the larger firms in the region closely cooperate with public research. For smaller firms, the situation is different. To smooth the cooperation between public research and SMEs, the intermediary organisation Culminatum plays a major role. The aim of this network organisation is to ameliorate the knowledge exchange between universities and SMEs and to act as business incubator (see also section 4.3). In addition, it is active in raising European funds for all kinds of research and technology projects. The organisation has five areas of expertise, one of which is telecommunication. It encourages technology transfer in several ways. First, it helps universities and polytechnics to become more aware of the needs of (small and medium-sized) firms. This has contributed to a more business-related research and education agenda in these institutes and increased the quality of education. The increasing contract research has strengthened the financial position of universities, whereas central government funds were increasingly cut. As universities and SMEs are very different culturally, and 'speak different languages', Culminatum acts as 'interpreter'. The project has also raised the awareness of universities and schools of their product. The organisation has limited means, with a budget of FM 10 million and 10 staff members. With these small means, Culminatum has succeeded in initiating many projects with a total value of over 100 million FM in the last three years. In particular, consderable funds were raised from the EU and national sources.

A weak element of Culminatum is that large firms in the region – such as Nokia and Sonera – are not directly involved in the technology transfer activities, although one Nokia representative has a seat on the steering group. A more operational involvement of big firms – which constitute enormous sources of knowledge and expertise – could help very much to strengthen both the quality of SMEs in the cluster and Nokia's strength. A hopeful sign is that Nokia seems to have shown an interest in the subcontractors' network which Culminatum plans to set up.

For the future, the ambition of Culminatum is to increase further the number of SMEs involved in R&D and university cooperation. In particular the involvement of non-high-tech firms is very low. The aim is to create a one-stop shopping network with one 'network point of entry', as an integral helpdesk for all kinds of SMEs seeking partners for innovation.

Spin-offs from the public knowledge infrastructure Although traditionally, entrepreneurship in Finland is not very highly esteemed, the 'birth rate' of firms in the cluster in the Helsinki region is considerable. Most firms in the telecommunications cluster are set up by young people. An important programme in this respect is the Spinno initiative (in which Culminatum participates, with others). Spinno offers specific training for academics who want to become entrepreneurs. In Finland, there proved to be a large demand for such service, partly because individual researchers, not universities, own intellectual property rights of their new inventions, which generates a great incentive to start up business. During the last five years 160 new businesses have got under way. In the Arabianranta area, there are spin-off firms from the Art and Design Academy. Their number is expected to increase with the completion of the art and design city project.

5.3 Linkages within the Public Knowledge Infrastructure

Because of rapid technological development, the field of telecommunication has long been almost entirely technologically driven. However, although technology remains important, other skills are gaining weight for many firms in the cluster. For instance, the design of high-tech products is a very important success factor in the market. Also, marketing (knowing the needs of the client) and organisational skills are essential in the very competitive telecommunication market. Increasingly, firms need people who combine several skills and are able to generate innovative integral solutions, not only in the technical sense. In this light, cooperation – both in education and research – between the several specialised universities in the region could benefit the telecom cluster very much, resulting, for instance, in broad, integrated education programmes. However, in the Helsinki metropolitan area strategic cooperation between universities is virtually absent.

The major problem is that the relationships between universities and the separate municipalities are not as good as they could be. For instance, the municipality of Espoo has a poor relationship with the HUT. An example might illustrate this: the HUT planned to construct an open, broadband campus network which would also be linked to students' accommodation. Such a high-grade network could stimulate the development of new services, and give Espoo an advantage in that respect. However, Espoo did not want to contribute to this project, as it was regarded as elitist. In Helsinki the municipality had almost no contact with universities in its territory until recently, let alone with the HUT in Espoo. With the investment of the municipality in the art and design city project, the situation has slightly improved. In other Finnish cites, notably Oulu and Tampere, the situation is completely different: municipality and universities jointly build up regional economic and technical excellence.

In our opinion, given spatial and functional complementarities, cooperation in the Helsinki urban area should be on a metropolitan level. The municipalities of Helsinki (location of the Art and Design Academy and Helsinki University) and Espoo (location of the Helsinki University of Technology) should urge universities to cooperate much more. The city of Tampere can again serve as an example: there, the municipality urged the two universities (one technical, one general, with a distance of 10 km between them) to cooperate closely and additionally invested in a science park in close proximity to the technical university to optimise research-business interaction. It is illustrative that in the last few years, research budgets (from public and private sources) of Tampere's technical university have increased faster than those of the Helsinki University of Technology.

The municipalities of Helsinki and Espoo are partially aware of the possible benefits of better cooperation between universities, but up to now they have considered the stimulation of university cooperation the task of the national government. It might, however, be preferable to take a strategic attitude, like Tampere's, and start on the inter-municipal (metropolitan) level. A first step could be to make a contribution to the total functional integration of the intranets of the Helsinki University of Technology on the one hand and the art and design city on the other, or at least construct a broadband link between the two networks. That would provide a first means (or platform) to improve the integration of the Helsinki University of Technology and the Academy of Art and Design.

6 Confrontation with the Framework

Having extensively described the actors and relations in the cluster, this section puts the telecom cluster in the Helsinki region in the general research framework developed in Chapter Two. The cluster is put in the spatial-economic context of the Helsinki metropolitan area. In addition, the different aspects of organising capacity regarding the cluster are discussed.

6.1 Spatial-economic Context

In the spatial and economic context in which the telecom cluster has its function, the following aspects can be discerned: the economic structure of

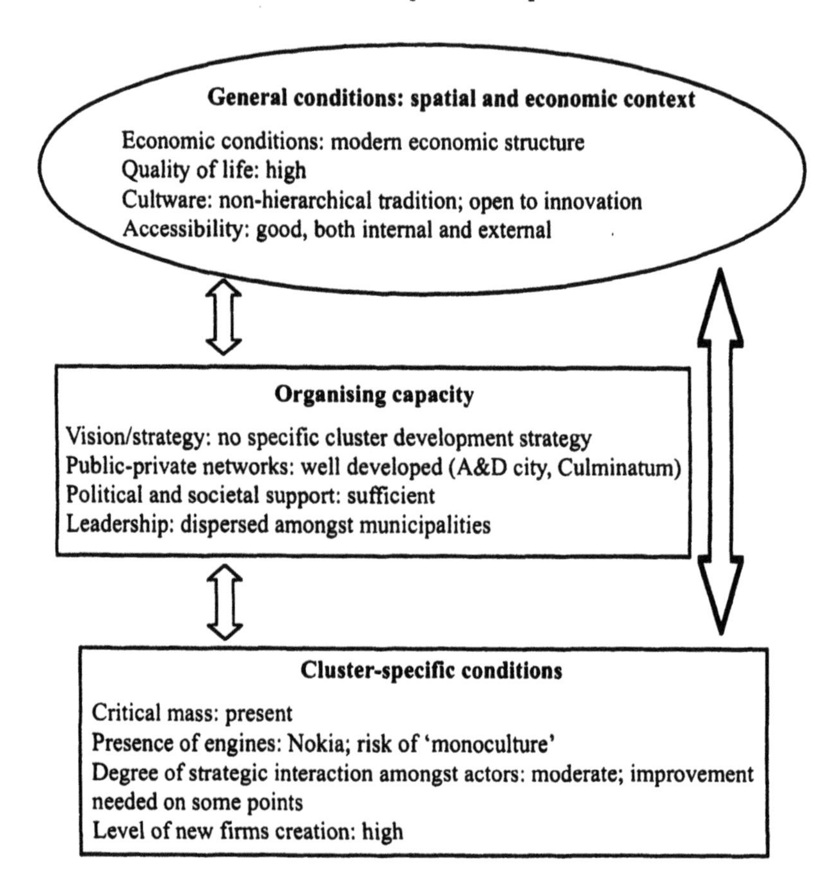

Figure 5.4 Framework of reference

the region, the cultware (i.e. cultural/societal aspects), the quality of life of the region and the accessibility of the region.

The *economy* of the Helsinki region is well developed and strong. After a period of severe decline, growth rates rank well above the European average. The economic structure of the region is diversified, with much high-grade industry and many business services. In particular the presence of a rich variety of financial and other business services is important for the functioning of the telecom cluster, but also the traditionally strong, advanced industrial sector was an fertile soil on which the cluster could flourish. The local demand for advanced telecom products and services is high in Helsinki. For instance, as the capital of Finland, Helsinki hosts many headquarters of Finnish large firms, which are important and demanding clients. The same holds for the government agencies in the city. The knowledge base in the Helsinki region is very well developed. People have a very high level of education on average, which is a clear advantage for the telecom cluster, in which advanced and well-educated labour is one of the key success factors. The presence of a number of universities contributes greatly to this. Moreover, Helsinki acts as a magnet for young talent from the countryside, so that the knowledge base in the region is still increasing. The region of Helsinki can be considered an important innovative milieu.

The cultural and social aspects of the Helsinki economy (the 'cultware') are very important to the success of the telecom cluster in Finland. Firstly, the non-hierarchical social structure – at least compared to countries such as the UK, France and Germany – facilitates the dynamics of interpersonal networks, which are very important in a turbulent sector in which so much change takes place every day. The resulting flexibility and short communication lines – both within and between firms and institutes – permit innovative and fast responses. For a large firm like Nokia, such aspects play an important role in the decision to keep its headquarters in Finland. Also, foreigners, in particular people from countries with strict hierarchical traditions, are attracted by these cultural aspects. Secondly, the Finnish people in general, and Helsinki people in particular, seem to be very keen to adopt new products and services. For instance, mobile telephone penetration has passed 50 per cent, the highest share in the world. Also, the adoption of computers and the Internet in Finland is faster than in most other European countries. A slight drawback of the 'cultware' of Helsinki firms, particularly when compared to Swedish competitors, is some degree of 'local thinking'. Whereas Swedish firms are historically very internationally oriented (for instance, Swedish new media firms are very fast to export), their Helsinki counterparts seem somewhat more conservative in this respect, and expand at a slower pace. Another drawback is the lack of private venture capital. This relates to the conservative financing tradition of banks and other private investors, compared to, for example, the USA. As a result, it is relatively difficult to start up new firms (which are important sources of dynamics in the cluster).

The *quality of life* in Helsinki can be considered good. The Helsinki region offers metropolitan amenities – theatres, opera, museum, nightlife – while at the same time the countryside is never far away. These characteristics, among others, make Helsinki a pleasant place to live. On the other hand, the climate – cold, and rainy weather, long and dark winters – partly offsets these advantages. In general, the quality of life in the Finnish capital attracts many young people from other parts of the country, who prefer the urban, relaxed way of living in Helsinki. This benefits the actors in the telecom industry, who see the human potential in the region increasing. The unfavourable climate may in some cases be a drawback to attracting foreign firms and employees to the region, although this is not generally considered a problem. Still, if the cluster continues to grow at its current pace, staff shortages will become more severe and measures will have to be taken. To attract people from Western Europe and the USA will be difficult, not only because of the climate but also because of the heavy tax burden in Finland. An option is to hire people from the adjacent regions in the Baltic.

Regarding *accessibility*, by European standards Helsinki is a very peripheral city. However, this does not seem to hamper the development of the telecommunications cluster in Finland. With its international airport, the Helsinki region is well connected to the rest of the world. This is very important for the internationally operating firms, such as Nokia and Finnet Group. Electronic accessibility – very important for the cluster – is excellent. Within Finland and the Helsinki region, road, rail and air connections are good. Compared to Europe, congestion is very limited, mainly because of the abundant space.

6.2 Cluster-specific Conditions

As described extensively in sections 4 and 5 the telecom cluster in the Helsinki region is a well-developed and dynamic cluster with all elements present: a demanding home market, manufacturers, service providers, a rich diversity of supporting business services and a high-quality public knowledge infrastructure. An important role is played by a few ‘engines’ in the cluster, notably Nokia, Sonera and Helsinki Telephone Corporation. With their extensive strategic networks with smaller firms in the region as well as with other multinational firms over the world, they generate much energy and create spin-offs. Perhaps a better coordination of efforts could enhance the twinning of large firms with smaller counterparts. Strategic interaction of firms with the well-represented public knowledge infrastructure is well developed. The large players in the region are very well aware of the assets that the R&D institutes constitute and behave accordingly. Similarly, the universities see the firms in the region as partners for research and cooperate with them in education. A weak point is the lack of cooperation between universities. This is a serious drawback, as the firms in the telecom cluster urgently need people with a broad range of skills. The municipalities of Espoo and Helsinki could play a positive role in the stimulation of inter-university cooperation, but the large firms, as main ‘clients’ of universities, may also contribute.

6.3 Organising Capacity

In the organising capacity related to the cluster, several aspects can be discerned: the presence of vision/strategy; the quality of public-private networks; the level of political and societal support for the development of the cluster; and leadership.

On the metropolitan level – which is the relevant level at which to study this cluster – there is no specific *strategy* aimed at the development of the telecommunications cluster. The municipalities of Helsinki and Espoo, where the principal firms and institutes that form part of the cluster are located, each have their own objectives and goals. There is no vision as to how to develop the cluster in a metropolitan perspective. Although up to the time of writing this has not yet had disastrous consequences, possible synergies are about to be missed. Examples are the construction of completely separate broadband networks in Otaniemi Technology Park (Espoo) and art and design city area (Helsinki). Integration of the networks would be better, and offer a unique chance to link high-tech with design and creativity. Other very relevant synergies could lie in a better coordination of university research and education of the institutes in Espoo (HUT) and Helsinki (Art and Design, University of Helsinki).

As there is no joint metropolitan strategy, there is no clear *leadership* in the cluster either. Leadership is clearly hampered by the dispersion of the cluster across Helsinki and Espoo. The private sector evidently takes the majority of initiatives, and does so very well. Still, a higher extent of public leadership could achieve much. The city of Tampere can serve as an example: the municipality took the initiative to build a high-tech region with all partners involved (business, education, the public sector), from a clear and broadly shared vision, and even managed to convince the two local universities to increase their mutual cooperation. On the other hand, leadership should also come from the private sector, notably Nokia, as by far the most important player in the region, to guide the organisation of regional networks.

In the cluster, the private and the public sector work together relatively well. Examples are the Culminatum programme, the Media Studio, the art and design city in Helsinki and the well-developed links between the large firms and public universities. However, these networks are dispersed, again mainly because of the two municipalities that are involved. There are, however, some signs of a change for the better. One such sign is the establishment of the Helsinki Club in autumn 1997, whose members are influential representatives from the business community, public administration, science, the media and the cultural community, as well as the mayors of Helsinki, Vantaa and Espoo. The objective of the club is to seek opportunities that benefit the Helsinki region as a whole, among other things by enhancing mutual trust amongst the actors, to remove bottlenecks and to coordinate the many activites in the area. One of the strategic areas is the preparation of a joint development strategy for scientific and art institutes, other university-level institutes and the associated entrepreneurial activity (Helsinki Club, 1997). In short, this club could achieve much in the cluster and may act as a driving force for new initiatives.

7 Conclusion and Perspectives

Owing to early market liberalisation, a rich tradition in high-technology engineering and a favourable 'network-friendly' culture, the region of Helsinki now hosts a telecommunication cluster of impressive scale, with all the necessary elements and high growth potential for the future. Engines behind future growth are the ongoing market liberalisation process in Europe, inducing lower prices, more diverse services and more demand, and the rapid technological developments which offer new opportunities – in particular in foreign markets – for the Helsinki-based telecommunication equipment and service providers.

Its international competitiveness is clearly derived from local strengths. A strong asset in Helsinki is that apart from the high level of capabilities of the individual players – firms (big and small), education, research universities – strategic interactions in the cluster are generally well developed as well, inducing a rapid transfer of new ideas, information and knowledge in the cluster, enhancing flexibility and, as a result, generating a high pace of innovation. Nevertheless, there are some points which need improvement:

- in general, organising capacity at the metropolitan level – the relevant level regarding the cluster – falls short in some respects. One of the main shortcomings is that there is no integral vision and strategy regarding telecom and new media on a metropolitan level. This hampers the dynamics of the cluster, as synergies and new combinations remain unused;
- the municipalities of Espoo – where the high-tech part of the cluster is located – and Helsinki, with its more content-oriented part, should together take the lead in exploiting latent synergies. Possibilities are better coordination and functional integration of Innopoli and Arabianrannta, and the stimulation of cooperation between the Helsinki University of

Technology and the Art and Design Academy. The concerted approach of Tampere and Oulu can serve as best practice. Additionally, it is important to improve the commitment and knowledge of responsible civil servants about the telecommunications and new media sectors;

- in the private sector, leadership is needed to enhance the technology and knowledge transfer from big firms to smaller ones. If the many innovative small firms could benefit more from the experience, knowledge and networks of the internationally-active large firms, their growth could be much stimulated and enhance the dynamics of the cluster as a whole. Nokia, as leader in the cluster, is the most appropriate actor to initiate this, but also more concerted action on a metropolitan level is also needed. Culminatum may play a positive role, with its broad participation of regional actors and experiences with technology and knowledge transfer. However, its budget is insufficient to achieve this. A 'merger' between the network of Culminatum and the money of TEKES, a fund to finance university/firm cooperation, could create a very powerful structure on the metropolitan scale.

The establishment of the Helsinki Club, with prominent members from the metropolitan region, both public and private, could be an important platform to discuss strategic issues regarding the development of the cluster, discover synergies, develop a vision, create mutual trust, and streamline actions of the several parties involved.

References

City of Helsinki Urban Facts (1998a), *Statistics 1998.*

City of Helsinki Urban Facts (1998b), *Urban Industries, Helsinki and the Helsinki region.*

City of Helsinki Information Management Centre (1996), *Helsinki quarterly, 3/1996.*

Financial Times (1998), 'Acquisitions: Technology mergers at record level', 29 July.

Finnet Group (1996), *The white book of telecommunications.*

Helsinki Club (1997), *Success Strategies and Partnership Projects for the Helsinki Region.*

Helsinki Metropolitan Area (1998), *Statistics 1998.*

Holstila, E. (1998), 'Finland', in L. van den Berg, E. Braun and J. van der Meer (eds), *National Urban Policies in the European Union: Responses to urban issues in the fifteen member states*, Ashgate, Aldershot.

HTC (1998), *Annual Report 1997.*

Media Studio (1997a), *Uusmediateollisuuden asiakkaat 1997.*

Media Studio (1997b), *The Finnish New Media Industry, 1997.*

Nokia (1998), *Annual Report 1997.*

NRC Handelsblad (1998), *Alliantie van BT en AT&T in telecom*, 27 July.

Discussion Partners

Mr E. Holstila, City of Helsinki Urban Facts, Managing Director.
Mr P. Kykkänen, Art and Design City ltd., Managing Director.
Mr O. Lindblad, SITRA Finnish National Fund for Research and Development, Director of Finance.
Mr K. Nordman, Helsinki Telephone Corporation, General Director.
Mr H.J. Perälä, Helsinki Chamber of Commerce, Managing Director.
Mr K. Ruoho, Culminatum Ltd., Managing Director.
Mr H. Saikkonen, Nokia Group and Professor of Telecommunication.
Mr I. Santala, Innopoli High Tech Center, Managing Director.
Mr N. Tuominen, City of Helsinki, Director of Business Services.
Mrs T. Varmavuo, Nokia Group, Head of Education.
Mr K. Weckström, Interaktiiven Satama, Managing Director.

Chapter Six

The Media Cluster in Leipzig

1 Introduction

Leipzig has long been a media city of considerable importance. Before World War II, the city was one of Germany's principal publishing centres. Furthermore, it was the major locus of the printing industry and manufacturing of printing equipment. After the war, Leipzig's position weakened considerably: in the early years of the German Democratic Republic (GDR), many publishers fled to West German cities (notably Frankfurt am Main, Munich, Nuremberg), and the influence-sphere of Leipzig's media industry remained restricted to the GDR. After 1989, with the fall of the Iron Curtain, and soon after German unification, Leipzig's position changed again. Despite severe difficulties, the situation offers new chances to build on traditions in a new way. Currently, the city of Leipzig aims at revitalising its media past by building a new media future, with an explicit policy to develop the media industry.

This chapter contains an analysis of the state of the art and the development potential of the media cluster in the city of Leipzig. A broad perspective is taken: not only the printing and publishing sectors are considered, but also the audiovisual media (television, radio), as well as new digital media (software, multimedia) and the Leipzig Fair.

This chapter is organised as follows. Section 2 describes the development and structure of the economy of the city of Leipzig, as Leipzig's media industry should not be studied in isolation but embedded in the local urban economy. In section 3, as a background, some brief developments are sketched in the media and communications industry in general. Section 4 presents the principal players in the media cluster in Leipzig, consisting of firms, education and research institutes, as well as local and regional governments. Section 5 analyses the dynamics of the cluster: it describes strategic relations between the several actors in the cluster, and the subject of new firms creation. Section 6 puts the cluster in the framework of reference that was developed in Chapter Two. Section 7 concludes.

2 Leipzig: Profile, Economic Structure and Development

2.1 Profile

The city of Leipzig is situated in the eastern part of Germany. It is part of the densely populated Free State of Saxonia. Currently, the city has 492,000 inhabitants. Since 1990, the population has decreased.

During this century, Leipzig's fortune and its position in Germany (and Europe) changed fundamentally several times. The most striking changes came with the German divide after World War II, and with the fall of the German Wall and German unification in 1989/90. In the first half of the twentieth century, Leipzig was one Germany's principal economic centres. Situated very centrally in Germany, at crossroads of rail and road, the city had an economic and cultural importance that reached far beyond its immediate surrounding area. By then, in terms of population too, the city was one of Germany's main centres. After the war, Leipzig became part of the socialist German Democratic Republic, and lost much of its economic and cultural importance. 1989 marked a new era again, when Leipzig became part of the unified Germany.

Some 10 years later, although much has been accomplished to revitalise the city, the position of Leipzig (as that of so many other ex-GDR cities) is still far from optimum in several respects: the city is still plagued by high unemployment rates, disappointing rates of growth and out-migration of (young) talent and skilled people to the wealthy western part of Germany. On the other hand, there are many signs of hope as well: cultural life in the city is very vivid, much has been improved in the environment, and at the time of writing, it is expected that the city will shortly be connected to the German high-speed rail-network, and the airport will be renovated.

2.2 Economic Development

The economic development of Leipzig during the last 10 years is characterised by transition from a centrally-led economy to a market economy. After 1989, the Leipzig economy became subject to the 'wind of capitalism', entailing fundamental economic changes. Old economic activities disappeared and new ones emerged at an astonishingly rapid pace. In industry, many jobs were lost. This loss was only partly compensated by a resurgence of the service sector. Still today, the Leipzig economy, like that of many other ex-GDR cities, is in its transformation stage.

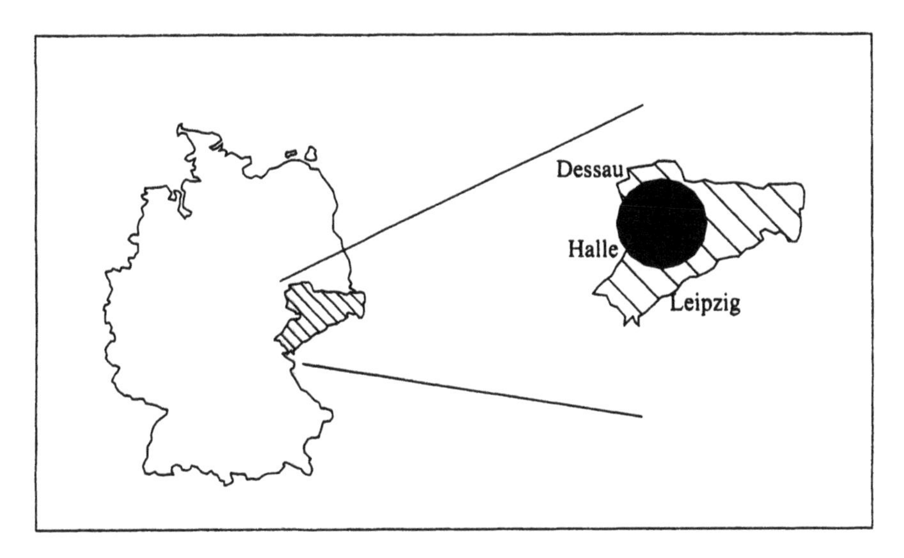

Figure 6.1 Germany, Saxony, Leipzig

Source: http://www.ris-ritts.epri.org/regions/halle.html.

The large scale industries and the energy/mining sectors were the most heavily hit by the changes. Many of the formerly state-owned industrial conglomerates were not able to compete with their Western counterparts and had to downsize considerably, or even close down. In particular machine building, one of Leipzig's specialisms, was decimated. The losses in industry (and also in the energy sector) were only partly compensated by a rise of the service sector. This was a 'catch-up' operation, since as almost everywhere in the ex-Eastern bloc countries, services had been underdeveloped under the production-oriented socialist system.

The changes of 1989 were on the whole accompanied by heavy job losses. Many unproductive jobs that had been kept up in GDR times became redundant and many factories closed down or downsized. Even in the last few years, industry has lost jobs. As in eastern Germany as a whole, unemployment is a major problem in Leipzig. From 1990 on, the number of unemployed people in Leipzig doubled, from some 23,000 in 1990 to over 45,000 in 1997, being 16.1 per cent of the labour force. These figures are slightly below those of the Free State of Saxony (17.1 per cent) and east Germany (18.1 per cent). In 1997, unemployment in the construction sector grew sharply to over 5,000 people, the main reason being the end of the 'construction boom' that Leipzig experienced during the first few years of unification.

2.3 Economic Structure

The structure of the economy of Leipzig is nowadays dominated by service activity (see Table 6.1). As stated earlier, much industrial activity has disappeared during the last decade. Within the remaining industry, relatively strongly represented sectors are the paper and printing industry (29 per cent of industrial firms), machine-building (21 per cent) and the food industry (15 per cent). In manufacturing, new jobs have been created as well: the city of Leipzig has managed to attract some new industrial investment from the West, such as a Siemens cellular phone plant (headquarters in Munich).

One characteristic of the industry is the dominance of small firms and low representation of large, internationally active firms. This is disadvantageous for the development of smaller industrial firms that are often dependent on orders from the industrial giants.

In the last decade, the service sector has grown at a rapid pace. Many trade firms, 80 banks, 59 insurance companies and many other business service providers settled in Leipzig, most of them subsidiaries of large German companies.

Table 6.1 Sector structure of the Leipzig economy, 1997

Sector	Employment, %
Agriculture	0.3
Energy/mining	13.7
Construction	12.1
Trade	11.6
Transport/communication	5.5
Banking/insurance	4.2
Business services	39.7
Non-profit	4.3
Other	8.6
Total	100

Source: Stadt Leipzig, 1998.

2.4 Revitalisation Policies in the City of Leipzig

There are many initiatives and stimulation measures aimed at regeneration of the weak economy of Leipzig. Firstly, Saxony (and all the other ex-GDR states) is a European Objective-1 region, and thus receives considerable financial contributions to enhance its economic structure. Secondly, to help ex-GDR cities and regions to cope with their transition process and to generate employment, there is a nationally and regionally supported programme named 'Gemeinschaftsaufgabe Verbesserung der regionalen Wirtschaftsstruktur' (common task amelioration of regional economic structure). This programme mainly promotes investments in the ex-GDR by providing investment subsidies. In the period from 1990 to 1997, 503 projects were subsidised, with a total contribution from the programme of DM 962 million (Stadt Leipzig, 1998).

The city of Leipzig also operates a number of economic development projects. Much of its effort is aimed at new firm creation. For instance, the city operates a fund (together with the Kreissparkasse Leipzig) worth DM 5 million, to provide venture capital on favourable conditions.

3 General Developments in Media and Communication

Before we turn to the description and analysis of the media cluster in the Leipzig region, this section puts its position in the context of the much broader communications industry. Very briefly, the most important trends and developments in this industry are discussed, as are the implications for the audiovisual industry in general.

3.1 Trends and Developments in Communications

The communications industry is a very dynamic part of the economies of the Western world. Developments in communications have been very fast and have played a fundamental role in the change of nature of Western economies into information economies or even information societies (Castells, 1996; Croteau and Hoynes, 1997). A rough distinction can be made between developments on the demand side (the demanders of communication services) and on the supply side. They influence each other, as will become clear in this section (see Figure 6.2).

Changes in demand for communication In general, the demand for communication services is increasing (Booz.Allen&Hamilton, 1998; Smits, 1998). This is most obvious in the business world. As (international) competition is growing, communication with clients is vital to achieve competitive advantage. In a straightforward way, communication is used to influence clients' decisions, by all kinds of advertising, with the intention of making people buy the advertised products ('one-to-many communication').

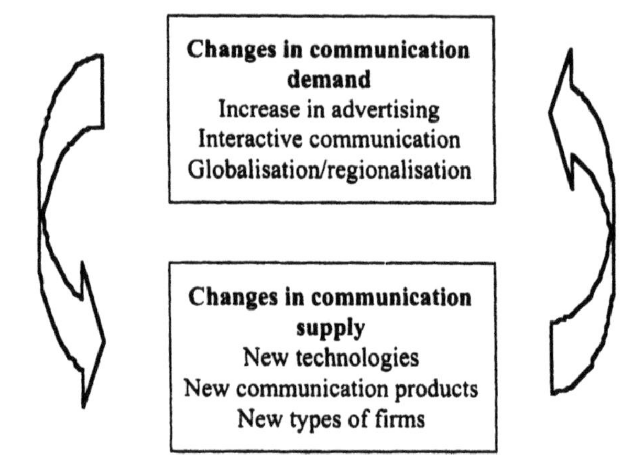

Figure 6.2 Demand and supply changes in communication

Communicating with clients is also increasingly used to understand and respond to market needs. Increasingly, firms try to communicate with clients in a direct way to understand their preferences and adapt products and services better to their needs, as a means to reach a leading competitive position. Examples of techniques and means used to achieve this are client cards and, more and more, such interactive media as the Internet. It is not only communication with clients that is important: the same holds for communication and interaction with all kinds of stakeholders: business partners, relations and personnel (Fidler, 1997).

Firms are not alone in demanding changes in communication: the same holds for consumers. The arrival of new media and communication techniques has changed the use of media and communication means. An important aspect of demand changes in communication is individualisation and differentiation. As an example, more and more tailor-made media products are available to individuals, such as pay-per-view and video-on-demand concepts. Digital techniques are making these services possible. In communication, the use of the Internet and other on-line services is increasing rapidly.

An important change in (mass) media is the trend towards regionalisation: people increasingly attach value to information about their direct environment (partly as a reaction to European integration and a globalising economy). On a national scale, this is reflected in the increasing importance of national television programmes in the national language instead of American series. This development has prompted US producers to seek entry to the European market by different methods, such as buying stakes in local producers, participating in joint ventures or making ad hoc deals with local producers (McKinsey, 1993). But at the same time, regionalisation is going even further. This is reflected in the growing importance of local and regional radio and TV stations in the last decade.

Changes in supply In the communications industry, very important changes occur on the supply side. Mainly under influence of digital techniques and possibilities of on-line communications, new products are offered and new types of firm enter the market. In the first place, many new products emerge and gain market share, sometimes very rapidly. Table 6.2 presents a set of new on-line products that are offered thanks to the emergence of new information and communication technologies. A distinction is made between different types of service: information services, interactive services, transaction services and entertainment services. Many new types of service have emerged in the last few years, in many different segments of the communications industry. The growth expectations of the different segments in the table are impressive. In the Netherlands for instance, yearly expenditures per capita on on-line communicating is expected to rise from DFl 80 to DFl 550 in the coming five years (Booz.Allen&Hamilton, 1998). In the near future, on-line services are expected to develop into a mass media and partly replace traditional media.

In the second place, the emergence of new services goes hand in hand with new market structures. New players enter the market, such as internet access providers, webpage designers, DTP studios, digital graphical firms, multimedia producers, CD-ROM producers, etc. At the same time, traditional (media) firms often extend their activities or incorporate new media in their product. Publishers set up an on-line publishing division, newspapers go on-line, bookstores sell on the Internet, banks introduce telebanking, and so on.

In the supply of the more traditional audiovisual media, such as broadcasting and video, there have been important changes as well. An important development concerning the audiovisual sector in the last decade in several European countries was the breaking of the monopoly position of the public broadcasters and the emergence of commercial ones. This has radically shaped the audiovisual landscape of these countries. Particularly importantly, the emergence of commercial television has entailed and reinforced the development and growth of production firms.

Table 6.2 Segmentation of on-line services

Communication between actors	**Information**	**Interaction**	**Transaction**	**Entertainment**
Expected growth in Europe (1998–2001)	31%	33%	75%	48%
Business-to-business	Product information On-line databases	Presentations E-mail Video-conferencing Direct marketing	EDI Telebanking Teleworikng	
Business-to-customer	Information Advertising Product information	Direct marketing Helpdesks Clients service	Telebanking Teleshopping E-services	Music TV/Video Pay-per-view Video-on-demand Games Edutainment
Government-to-business	Statistics Information services	E-mail Chatting Newsgroups	EDI for contracts Tax forms	
Government-to-consumer	Information		Tax forms Elections Permissions	Edutainment Lotteries Travel
Consumer-to-consumer	Homepages Advertisements			Multiplayer games

Source: Booz.Allen&Hamilton, 1998.

3.2 *The New Position of the Audiovisual Industry in the Communication Industry*

In general, activities in the traditional audio visual industry are directed towards the creation and production of feature films, presentation movies for firms, informative/educational productions, commercials, cultural films, productions of broadcasters, and animations (NEI, 1994). Several types of actor can be discerned: principals, producers, facilitators, freelancers and service providers.

Although many traditional audiovisual products will continue to exist, the rapid development in new media, in particular on-line media, and the emergence of digital technologies will change the way audiovisual products are used and incorporated. First, there is the digitalisation of 'traditional' audiovisual activities and second, the new information possibilities invoke new products and services. Examples of the digitalisation of 'traditional activities' in the media branch are films and videos made with the aid of computer programs and digital techniques, resulting in a better quality of sound and vision.

Perhaps more importantly, the new information technologies and infrastructures entail new products and services. There are many examples: pay-per-view, interactive television, information services for car drivers, music and video on demand services. New interactive media also provide opportunities for audiovisual products: for instance, sound and vision elements are more and more incorporated in internet sites and CD-ROMs. With the growing need for many different types of communication, this offers great scope for growth potential in audiovisual media, as sound and images are important elements in many (on-line) communication forms. Although in a different context, the production of these 'new' audiovisual products generally requires the same artistic, technical and commercial skills as that of 'traditional' audiovisual productions.

In sum, it seems reasonable to state that there is great scope for media cluster development. In the first place, the role of traditional audiovisual products such as films, videos and TV productions remains more or less constant. At the same time, the demand for communication is increasing rapidly, because increasingly firms derive competitive advantage from good communications strategies. Here, much growth can be expected, in particular in the incorporation of audiovisual products in new media. Audiovisual or related products (sound, vision, moving images) are progressively becoming integral parts of on-line concepts such as websites.

4 Actors in the Media Cluster in Leipzig

4.1 Introduction

This section describes the principal actors in the media cluster of Leipzig. In our cluster perspective, not only private media firms are important: education and research institutes are also taken into consideration. Furthermore, media is defined in a relatively broad sense: not only are traditional media branches (such as publishing and radio/TV/film) described, but also new digital media. Finally, the Leipzig Fair will be introduced. Though strictly speaking it may not belong to the media cluster, it has great importance as symbol of the city of Leipzig's tradition, as a major communication instrument and as strongly interlinked with media firms.

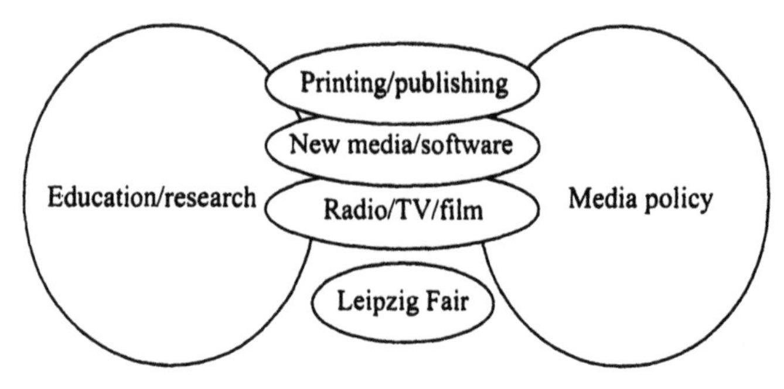

Figure 6.3 The media field in Leipzig

Table 6.3 Number of firms, employment and turnover in Leipzig's media cluster, 1996

	Number of firms	Employment	Turnover (million DM)
Publishers	118	2,376	291
Press	40	6,501	963
Radio/TV/film	65	1,386	168
Graphical industry	234	5,610	366
Software/multimedia	56	2,541	333
Fairs/exhibitions	18	2,409	321
Education/research	21	2,673	69
Total	556	23,496	2,511

Source: Freistaat Sachsen, 1997, adapted.

4.2 Media Firms/Institutes

Printing and publishing Before World War II, Leipzig was the publishing capital of Germany, with some 600 publishers located in the city. Linked to this, there was a sizeable printing industry. The relics of this past can be found in the city's Grafisches Viertel (Graphical Quarter). The war and the communist period destroyed Leipzig's position, as most publishers moved to West German cities of Munich, Nürnberg and Frankfurt. At the time of writing, the number of publishers has decreased to 118, many of which are small agencies of the great German publishers. One large player that has survived is Reclam. Although its headquarters are in Stuttgart, it has a large Leipzig location. Between 1990 and 1996 employment in publishing decreased considerably. Currently, Leipzig holds a relatively strong position only in a few niche markets (notably 'belles lettres', some science fields, language and local issues). Another interesting niche is electronic publishing, though growth in this segment has not met earlier expectations. It still seems to offer potential for growth, mainly because of Leipzig's superb electronic infrastructure and skilled people. For instance Klett (a big player in vocabularies) aims to set up its department for electronic publishing in Leipzig.

A positive development for the publishing sector in the city is the success of the Buchmesse. Instead of being a trading fair (like Frankfurt), the Leipzig Book Fair aims for direct contacts between authors and readers and manages to draw an increasing stream of visitors.

In connection to the printing and publishing sector, Leipzig was a major centre of graphical machine building before World War II. Even during GDR times, this position remained strong. However, after the fall of the Berlin Wall the machine-building sector in Leipzig almost disappeared, under severe competition from West Germany. Only a few manufacturers and technician bureaux have survived.

Radio, television, film In the field of radio and television activities, the main player in the region is MDR (Mitteldeutscher Rundfunk). It is a member of the ARD (Allgemeine Rundfunk Deutschland) family, and thus obliged to produce a certain percentage of programmes for ARD. MDR fulfils its broadcasting function for the states of Saxony, Thüringen and Sachsen-Anhalt. MDR's department in Leipzig operates a number of radio stations, as well as a TV programme for the Free State of Saxony. The major part of MDR's TV activities currently take place in Dresden. However, in a treaty, the states of Saxony, Thüringen and Sachsen-Anhalt arranged to concentrate all TV-activities in Leipzig. Before 2001, the TV department of MDR will move all its studios from Dresden to Leipzig. New studios are being built (value: DM 16 million) in the city. On that spot, space will be created for other media companies as well (see section 4.4 on the media-city project).

Leipzig has many radio stations, both public and private. The most innovative private radio station is PSR (Privater Sachsischer Rundfunk), one of the largest commercial radio stations in Germany. PSR uses broadcasting techniques in a number of new and promising areas, such as in traffic guidance systems, advertisements and tourist information. For instance, they have developed a multilingual programme that tells people through a headphone what they see during tourist city walks. In developing new products, PSR is keen to cooperate with local partners such as software companies and Deutsche Telekom. The company employs over 100 people (Stadt Leipzig, 1997).

Although the city does not have a real tradition in television production as it has in publishing, Leipzig hosts a few audiovisual production firms, mostly working for MDR. The largest is Saxonia Media, founded in Dresden, 1995 and located in Leipzig since 1996. The firms produces 1,000 minutes of fiction a year, mostly for MDR television, including the well known series 'Tatort'. In addition, there are many small audiovisual production firms that work for the business-to-business market (company films, marketing). Leipzig hopes to attract more TV production firms once MDR has moved its studios to Leipzig. In the media-city area, studios and offices are being built to host production firms that will work for MDR. Currently, it is difficult for (TV) production firms like Saxonia Media to attract people (actors, directors, camera-people, producers, location managers etc.): the concentration of film and TV production in Berlin in GDR times has drawn much talent from the region. Saxonia Media has solved this problem by educating local talent (from theatres) and 'importing' knowledge from Bavaria (the director of Saxonia Media was formerly director of Bavaria Studios).

Film production is scarce in Leipzig, although this situation may change as Saxonia Media plans to become active in cinema production. The recently established media fund of the Free State of Saxonia may also enhance film production. An increase of film-making in the city may be important for the cluster, as it may help to attract artists and other staff and help to create critical mass for technical facilities that are also used by television production.

New media/software The final sub-sector in the media sector consists of new media and software. This sub-sector is very important for the media cluster, because it has links to the other media sub-sectors, printing/publishing and television: digital techniques, data management/transmission and new communication technology are gaining importance in the media sector as a whole.

Between 1990 and 1996, employment in software and multimedia grew considerably (Freistaat Sachsen, 1997). The largest players are branch plants of global players in the field of software and ICT (Siemens, IBM). There is also a successful and fast growing Leipzig-based computer manufacturer, PC-Ware. There are a number of small firms active in data processing and software applications of all kind. Most of them are small, with fewer than 10 employees. A major advantage of Leipzig is its excellent electronic infrastructure. Since 1989 much effort has been put in upgrading the outdated and insufficient heritage from the past. New buildings – such as the fair and the Medienhof – are very well equipped, thus offering major advantages for their users.

Nevertheless, the number of firms active in new (multi)media and software is low compared to other German cities. There seem to be three basic reasons for its backward position. First, during GDR times the sector lagged behind West German cities: the level of technical development and sophistication was much lower and the infrastructure was poorly developed. Second, it has proved difficult to build up the sector because of labour market shortages in the field: graduates from university and polytechnics leave for west Germany, where salaries and career possibilities tend to be higher. Third, the development is hampered by a weak local demand for new media and software application, mainly due to the weak economic structure.

The Leipzig Fair An important trading centre, Leipzig has an impressive tradition as city of fairs which spans over 800 years. From 1894 on, large-scale exhibitions took place to display new brands and sample stocks. This concept was invented in Leipzig. The modern trade fair was born and grew very fast. For instance, Leipzig had the largest book fair in the world, strongly connected to the publishers in the city. During the communist period Leipzig continued to function as a trade fair, albeit on a much more limited scale. The fair mainly functioned as showcase for communism and lost much of its economic importance (Rodekamp, 1997).

After the fall of the Berlin Wall, the question arose of how to revitalise the Leipzig Fair. In 1994, the decision was made to build a new fair on the site of a former airport, near to the highway and the airport and with good connections to the city. In 1996, the new fair was constructed, with four large halls, a total area of 102,000 m^2, and equipped with the latest technical devices (local electronic highways, multimedia, lightning) and services (restaurants, stand design, and much more). Total investments amounted to an impressive

DM 1.3 billion. The fair is owned and operated by a daughter company of the city of Leipzig.

The new Leipzig Fair attracts some 1 million visitors and 12,000 exhibitors annually. Of the visitors, 60 per cent come from the ex-GDR states, 20–25 per cent from the rest of Germany, and 15–20 per cent from other countries. The number of exhibitors is considered too small. The main reason seems to be the weak economic situation in Leipzig and in the ex-GDR as a whole. An important drawback for the fair is the lack of international connections. Leipzig Airport offers few direct international destinations, as there is insufficient demand (again because of the weak economic situation of Leipzig). In the future the situation will improve, with the extension of the airport (to be ready in 2001) and the opening of a high-speed rail link running from Berlin to Munich via Leipzig (planned for 2002).

The Leipzig Fair has to be very innovative in developing new concepts and attracting visitors, as competition with other fairs (mainly those in Cologne, Hannover, Munich) is strong. A great success is the Leipzig Book Fair, aimed at personal contact between authors and their readers (in contrast to the Frankfurt Fair, which is oriented to book traders). Recently the Fair has become active in developing new products such as electronic exhibition (the Internet as 'virtual square metre') including face-to-face contacts between seller and vendor, made possible by video-conferencing. The Fair's superb electronic infrastructure gives it a lead over other fairs.

The Leipzig Fair and the media cluster are strongly interlinked: the Fair uses media for marketing and acts as a client for media of all sorts, but simultaneously the media (particularly TV and radio) use the Fair as an interesting location for live broadcasting. Section 5.2 will deal more extensively with the interrelations between the Leipzig Fair and the media cluster.

4.3 Media and Related Education in the Region

The city of Leipzig hosts a rich variety of education and research institutes active in the field of media. Table 6.4 presents the main institutes. For each media sub-sector, education facilities are present. Education related to the printing and publishing sector has always been strong in Leipzig, because of the city's position as publishing capital. Nowadays, the Polytechnic and the School for Graphic Design and Book-art are the main institutes.

A broad range of institutes at different levels offer education in audiovisual skills (radio, TV). At the university level, there is the Institute for Communication and Media Science. At a lower level, the TV Academy of

Table 6.4 Education institutes in media, Leipzig

Institute	Type of education	Number of students*
Hochschule für Grafik und Buchkunst (Academy of Art and Book Design)	Media-art, book-art, graphic design, new media	305
Hochschule für Technikwirtschaft und Kultur Leipzig (College of Engineering, Business and Arts)	Engineering, bookselling, publishing, library-management	4,078
Media Akademie	Education in media, computer design, multimedia	
Media Academy Leipzig Private School for Press, Radio, TV and PR	TV and radio production, PR, press services	
Fernseh Akademie Mitteldeutschland eV	AV production, journalism, management	
Deutsche Telekom AG Fachhochschule Leipzig	Telecom, news-electronics	364
University Leipzig, Economics Faculty, Institute for Software and System Development	Research/education in database, software development, teleinformation and multimedia service development	
University Leipzig, Institute for Communication and Media Sciences	Research/education in communication and media	

Source: Stadt Leipzig, 1997; Leipzig City Council, 1996; * Leipzig City Council, 1995.

Central Germany, the Media Academy and the Media Academy Leipzig Private School for Press, Radio, TV and PR can be mentioned. In the field of software and new media, there are courses at the University of Leipzig and at the Polytechnic. Furthermore, the city hosts a private school of German Telecom (Deutsche Telekom), the former German telecom monopolist.

4.4 Public Policies

The city of Leipzig has designated the media sector as a spearhead sector in its economic revitalisation strategies. This is not just an intention: there are many concrete initiatives. Further, the Free State of Saxonia plays an important role in the development of media, in particular in TV and production and film policies. This section gives an overview of the major policies and initiatives.

Media city Just outside Leipzig's city centre, the so called 'media-city' is under construction, near the future location of MDR's headquarters and studios. The media-city will consist of three studios and 15,000 m^2 of office space, as well as space for workshops (for instance, for set construction) and a media garden. The media-city is to become a site for TV and media production firms that work for the MDR station: MDR is obliged to outsource 55 per cent of its production. The project is due to be completed in the year 2000. The project is realised by the LGH (Leipziger Gewerbehofgesellschaft mbH) and DREFA. The LGH is a development corporation owned by the city of Leipzig, the Handelskammer and the Handwerkskammer. They own the land and commission the building activities. The project management is in the hands of DREFA (a subsidiary of MDR).

Medienhof The Medienhof is a restored, historic building in which many small and medium-sized media firms are located. The LGH (Leipziger Gewerbehofgesellschaft mbH) acquired the building in 1994, after it had become clear that media should be a spearhead in Leipzig's economic development, and renovated and reconstructed the building for some DM 20 million. Twenty-five media firms, varying from printing firms to multimedia studios use the 6,500 m^2. Ninety-eight per cent is occupied, a high rate by Leipzig standards.

Central German Media Stimulation (Mitteldeutschen Medienförderung) Since 1998, the States of Saxony, Sachsen-Anhalt and Thüringen have operated a fund to stimulate media productions in the Free State of Saxony. Saxony took the initiative. The aim is to stimulate the media structure in the region. Media activities/projects of all kind can apply for grants (film production (cinema and TV), as well as multimedia) but applicants are obliged to spend 100 per cent of this money locally. Stimulation of film production has priority. The long term objective is to increase the number of film producers in the region. The fund has a budget of DM 25 million (see Table 6.5 for contributors). Grants are assigned by a network of experts. Leipzig was chosen as location of the Medienförderung because of the presence of MDR and production firms. Similar funds in Germany operate in Bavaria (DM 60 million), Berlin, NorthRhine-Westfalia and Düsseldorf.

Stimulation of new media and software development An important initiative for software/multimedia development is the Leipzig-based institution named Saxony Telematics Development Corporation. This economic development corporation serves three goals: it sells Saxony abroad as a location for investment, particularly for IT and new media firms. Second, it aims at increasing consciousness and awareness of IT possibilities and it helps (small) IT firms

Table 6.5 Contribution of participants in media stimulation

Participant	Contribution (DM, million)
Free State of Saxonia	7.5
Sachsen-Anhalt	5.0
Thüringen	5.0
MittelDeutscher Rundfunk (MDR)	5.0
Total	25.0

Source: Interview.

in selling their products. For instance, they can get a grant to have a stand at the CEBIT Fair, a major computer and software fair in Germany. One of the recent activities of the corporation is a business plan competition amongst students. Winners receive all kinds of support to set up their own firms.

Also, the city of Leipzig is building a Business Innovation Centre (BIC), in which young entrepreneurs in the field of software and multimedia can be housed. The centre is developed in close cooperation with the university and polytechnic.

5 Interaction and Dynamics in the Cluster

This section analyses networks in the media cluster. In a dynamic and fast changing environment such as the media cluster, rapid and frequent interaction and cooperation between the different actors in the cluster is of great value: a well functioning cluster enhances flexibility, permits a better use of available knowledge, and boosts creativity in the region. Well-functioning strategic networks are likely to benefit all the parties, as they may generate advantages for the region as a whole. In Figure 6.4 the relations that will be described are schematically represented. The numbers in the figure correspond to the numbers of the subsections.

5.1 Interfirm Cooperation

Strategic interfirm cooperation in the media cluster is frequent, and takes many different forms. In the following, some examples will be given.

The audiovisual industry in Leipzig is not a complete sector in which all elements are present on a sufficient scale. In particular there is a lack of equipment/facilities firms in the region. This is an important impediment for

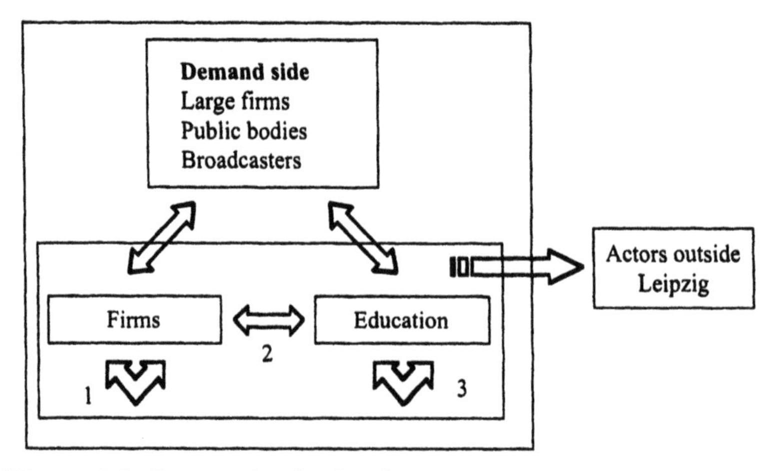

Figure 6.4 Interaction in the cluster

many Leipzig-based audiovisual production firms. In that respect, the firms are very dependent on facility firms in other parts of the country. For example, the most modern cameras have to be 'imported' from Berlin, which is relatively costly. The general feeling is that the scale of the sector is currently too small to support a sufficient facilities infrastructure. From the interviews, long-running productions (such as TV series) appeared important to attract facility providers, as well as the necessary actors, producers, directors, and technicians to the city: long term contracts provide security for staff.

In the audiovisual sector, particularly in TV productions, flexible cooperative production techniques are widely applied, as productions are made on a project basis. Saxonia Media, the largest TV producer in the region, has a permanent staff of only eight. Many people are hired on a flexible basis and many activities – lightning, camera assistance, sound assistance – are outsourced. The main broadcaster in the region, MDR, is obliged to outsource 55 per cent of its yearly production. The construction of the media-city is aimed at keeping the outsourced jobs in the region.

The radio stations in the city are more self-sufficient and less oriented to external cooperation, with the important exception of the very innovative PSR (Privater Sachsischer Rundfunk). They have strategic partnerships with software and IT firms – among which Deutsche Telekom – to develop innovative projects in which radio-transmission is used for new purposes. In the application of new services (traffic management, tourist guidance), they work together closely with the city hall.

In publishing, strategic interfirm cooperation on a local level seems scarcer. This industry is characterised by huge concentration on the world scale (with big players such as Bertelsmann, Springer and the Murdoch group 'eating up' many smaller firms). In electronic publishing, interesting partnerships of IT firms and publishers sometimes emerge.

Strategic interfirm relations in the multimedia/software sector are rare. This has to do with the specificity of applications: software and multimedia firms often offer one or a few innovative solutions, and try to sell these to clients. This sector is characterised by a certain degree of individualism.

The Leipzig Fair and the media The Leipzig Fair is connected to the media cluster in several ways. First, the Leipzig Trade Fair organises fairs that are of interest for the media cluster: the Book Fair and several fairs on electronics and software (CEBIT; BiK). Second, the Leipzig Fair acts as a client for the local media cluster: it uses TV and radio stations and the local press for advertisements (remind that 60 per cent of the Fair visitors come from the former GDR).

More strategic cooperation takes place when the Fair and a radio or TV station engage in partnerships. The Fair may offer opportunities for a live broadcast, for instance by providing studio space. This means 'free publicity' for the Fair (and helps to increase the number of visitor and exhibitors). For the radio and TV stations, to broadcast from a fair (most fairs are lively events) is also attractive. More generally, the Fair tries to link up with the city of Leipzig as much as possible. For instance, the opening ceremony of some fairs is held in the famous music hall in Leipzig's city centre (Gewandhaus). Furthermore, the attractiveness of the Fair depends on the attractiveness of the city.

5.2 Links between Firms and Education/Research

Media firms are linked up with education and research in several ways. First, most media firms benefit from the presence of much and varied education: they frequently employ students as trainees and benefit from the flow of graduates as a source of new staff. This holds for publishing, audiovisual media and new media/software (Freistaat Sachsen, 1997).

Audiovisual firms in the region complain about the lack of coordination between education institutes. In addition, the city lacks a college-level education for television and film production. Next, the education programmes seem insufficiently tailored to market needs. Therefore, it would be advisable to increase the influence of firms of the media sector on the composition of the education programmes. Furthermore, there might be a role for the state, as an important decision-maker in both education and the media sector.

Allianz Leipzig: where tradition and innovation meet

Allianz is one of the largest insurance firms in Germany. The firm was founded in Leipzig (Leipzig was the 'insurance capital' of Germany for many years), but moved its headquarters to Stuttgart at the time of the division of Germany. After the fall of the Iron Curtain, however, Allianz re-established the ties with its birthplace, partly for nostalgic reasons: it began to support the university of Leipzig in insurance-related education – mainly insurance informatics – and research, and to offer opportunities for students to have traineeships in Stuttgart. In this project, Leipzig University research staff and students developed a very innovative classification system for incoming mail for Allianz. As Allianz receives over 30,000 letters monthly, this invention is very relevant for the company. Allianz then decided to set up a special unit in Leipzig to put the system into practice.

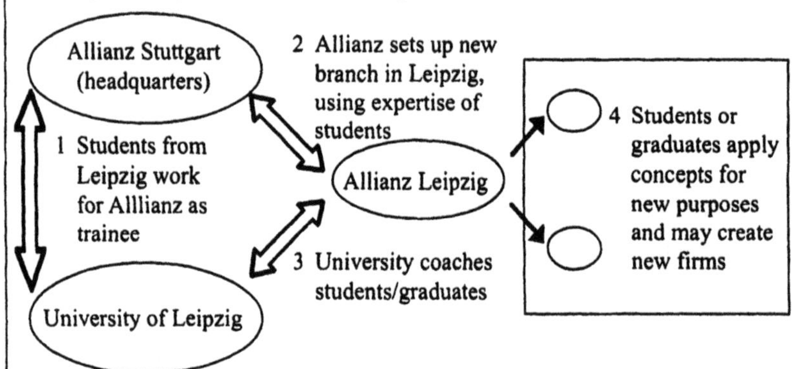

The main reason for the move was that graduates from Leipzig (ex-trainees for Allianz) wanted to stay in the city but had difficulties finding employment in the city, owing to the bad general economic situation. Interestingly, the technology used in the letter classification system can be applied in many other fields (media among them) as well. Thus, spin-offs may emerge from this unit, generating new high-grade employment in the city of Leipzig.

In software and multimedia, some firms are active in projects with the university. The Business Innovation Centre – an incubator for software and multimedia firms – was constructed in cooperation with polytechnics and university. An interesting and innovative cooperation model between business and university is sketched in the text box. It shows how new activities can emerge form the re-establishment of old relations.

5.3 Links amongst Education Institutes

Strategic cooperation amongst education facilities is important to optimise and coordinate the supply of media-related education in the city, thus making the city attractive for students who aspire to a career in media. Despite the variety and scale of media education in Leipzig (see section 4.2), the institutes' programmes seem to be insufficiently coordinated and not always tailored to market needs. Recently, MDR took the initiative to improve inter-institutional coordination.

On a university level, there is a far-reaching cooperation between the University of Leipzig and the university in the neighbouring city of Halle. They recognise each other's education programmes and coordinate their research efforts. Furthermore, together with the municipalities of Halle and Leipzig, they are engaged in the development of a single electronic student card system.

5.4 Dynamics: New Firm Creation

The creation of new firms is fundamental for the dynamics of the cluster as new media firms have opportunities to develop and in the long run may reach high turnovers and generate new employment. They may also be strategically important to the larger firms already present, which may benefit from the presence of small-scale media service providers in their vicinity. Since 1990, many media firms have been founded in the city: two-thirds of all media firms in Leipzig were created after that time, most in 1990 and 1991; 80 per cent of them were founded in Leipzig (Freistaat Sachsen, 1997). After this 'catch-up', the pace slowed: in 1996, only 11 new firms/institutes were set up.

For Leipzig even more than for other cities, new firms creation is important. The strategy of the city of Leipzig bears the name 'Leipzig, Stadt der Unternehmenden' (Leipzig, City of Entrepreneurs). Entrepreneurship of citizens is regarded as the key to revitalising and developing the hard hit economy. The university (and colleges) in particular are considered key assets and important sources of new entrepreneurs.

A number of initiatives have been taken to enhance firm creation. Examples are the Business Innovation Centre, the Medienhof, the business plan competition of the SET, the organisation of seminars for new businesses, and many others. However, despite great ambitions and efforts of the municipality, the 'entrepreneurial spirit' at the universities seems to be low. Among other things, this is reflected in the low number of spin-offs from the university.

Several reasons can be put forward. First, university education is highly theoretical. In most institutes, students are not trained in entrepreneurial skills: only the Technical University has a chair on starting a business. Second, students seem to be risk-averse: they prefer a permanent and secure job in a (large) firm to starting their own business.

On the other hand, too much effort aimed at new firm creation may have an adverse impact on the development of the cluster. We found that existing (small) firms, particularly those active in software/multimedia, have great difficulty finding qualified staff. One of the problems indicated was that the City Hall's (and the Free State's) policy to stimulate new business creation may hamper the growth of existing firms with expansion plans: if every talented graduate started their own firm, the fragmentation of the sector would be aggravated. To avoid such adverse effects, it might be advisable to supplement start-up stimulation policies with policies aimed at connecting university students and graduates with small, fast-growing firms.

6 Confrontation with the Framework

6.1 Introduction

In this section, the media cluster is confronted with the framework of reference of Chapter Two. This framework puts the cluster in the perspective of the urban context of the city as a whole (urban economic structure and spatial conditions) and analyses the organising capacity with respect to the cluster.

6.2 General Conditions: the Economic, Spatial and Cultural Context

With low growth rates, high unemployment figures and limited investments, the *economic conditions* in Leipzig are worrisome still. The relatively weak economic situation entails a lack of demand, and slows down the growth of business-related parts of the media cluster, particularly new media and software development, but also printing/publishing. For other parts of the cluster (notably the audiovisual industry) the economic development is less important as a driver behind sector growth.

Regarding the spatial conditions, the most important aspect is that the external *accessibility* of the Leipzig region is moderate but will improve in due time. The region is well connected to important centres by road, but not by high-speed rail. In addition, the number of European destinations from the

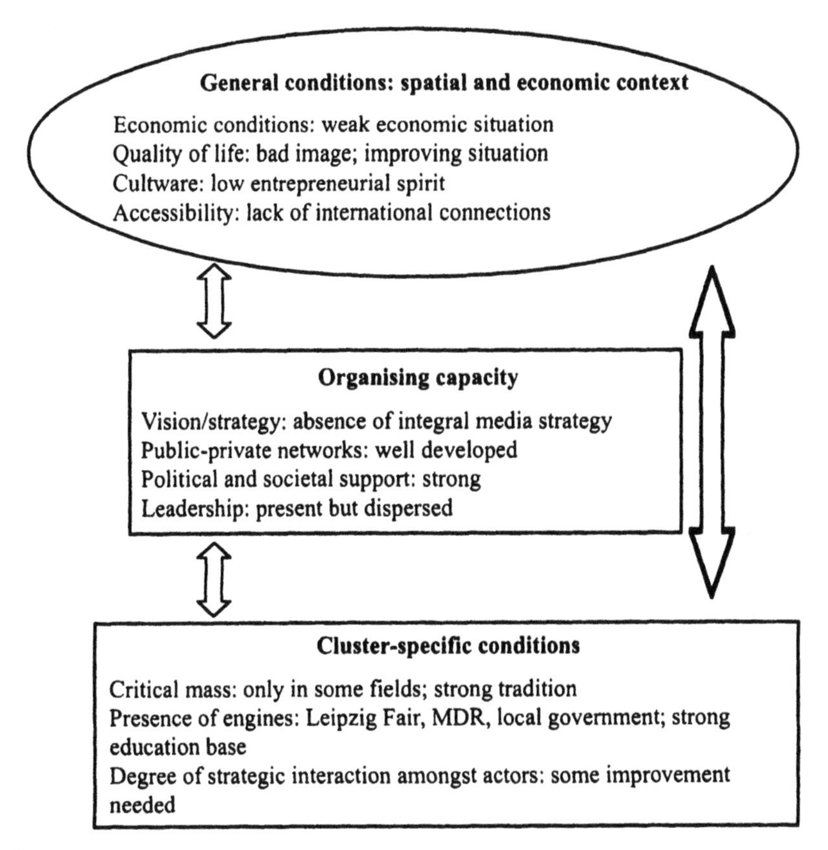

Figure 6.5 Framework of reference

airport is limited. This might be said to hamper the development of the Leipzig Fair and reduce the attraction of foreign investment in general. The 'electronic accessibility' of the city, on the other hand, is excellent, which entails advantages for the settlement of new media business.

Yet another important spatial condition is the *quality of life* and the living environment in the city. Much needs to be done in that respect to counter the negative image of Leipzig, as the key to attract (or retain) skilled people to the region. However, this seems to be changing: environmental improvements have contributed much to the attractiveness of the city. The availability of cultural facilities and nightlife are also judged positively. Very importantly, young people (students) generally appreciate the city's atmosphere and want to stay if there are job opportunities.

Cultware relates to the attitudes of people and firms, towards entrepreneurship, innovation and cooperation. For Leipzig, the cultware offers a diverse picture. For one thing, the entrepreneurial spirit seems to be relatively low. This is probably due in part to long years of centrally-led economy in which entrepreneurship was not rewarded (or even not allowed). But even in the 1990s, students did not seem to be very entrepreneurial: most of them preferred solid, permanent jobs. The willingness to cooperate in the cluster seems to be greater. The adverse economic conditions and the need to compete with other cities brings people together.

6.3 Strength of the Cluster

The strength of the cluster and the interaction between the elements has been extensively discussed in sections 4 and 5. To summarise, an important asset of Leipzig is its great media tradition. Recently, the audiovisual sector has developed strongly under the influence of MDR's relocation and its outsourcing strategy and strong stimulation policies. Publishing remains relatively weak, with growth opportunities in some niche markets. In new media and software, the dynamics are strong but the scale of the firms is small; cooperation is rare and development is hindered by the weak economy and the lack of skilled staff. Education facilities in the field of media are relatively abundant in the region, considering the very small size of the sector. However, cooperation among and between institutes and the cluster could be improved in several respects.

An important aspect of the spatial conditions in the cluster itself is the development of a few 'media locations' (the media centre as a focal point for the audiovisual cluster, the Medienhof, the BIC). The risk of competition among these centres should be minimised.

6.4 Organising Capacity

Several aspects of the organising capacity related to the audiovisual cluster can be discerned: the presence of vision/strategy; the presence and quality of public-private networks; the level of political and societal support for the development of the cluster; and leadership.

Although media is explicitly recognised by the city as a potential engine for development and many initiatives are taken (see section 4), there is no clear, balanced and elaborated *vision and strategy* yet on the development potential of the cluster as a whole: efforts are directed to parts of the cluster. It seems that media is considered too much as an independent, separate sector. More stress on its embeddedness in Leipzig's economy as a whole might lead to new and better insights. *Public-private cooperation* seems to be fruitful in the media cluster in Leipzig, as can be seen in the development of the ambitious media-city project. *Political (and societal) support* for cluster development is abundant, media being a priority sector to develop. Thus an important precondition for cluster development is fulfilled. *Leadership* in the cluster, finally, is dispersed among several parties, mainly the city of Leipzig and the Free State of Saxony (initiating projects, implementing stimulation measures) and the MDR broadcasting organisation (in organising networks with suppliers). The dispersion of leadership does not seem to generate difficulties.

7 Conclusions and Perspectives

The city of Leipzig has several assets that may be helpful in extending its position as media city. Perhaps the most important asset is Leipzig's great media tradition. This has become clear already with the commitment of formerly Leipzig-based firms like Allianz Insurances and publishers Reclam, which have opened subsidiaries partly for nostalgic reasons. The new Leipzig Fair is another great example of building something new on the past. Apart from being a tourist attraction in its own right and a city marketing instrument, the Leipzig Fair can play a role in the local media cluster as large demander of various media products.

Although a golden past and traditions are very valuable and unique, they are not sufficient to build a new future. Fortunately, Leipzig has more to offer. A major asset is Leipzig's strong position in radio and television broadcasting and production. The move of MDR, the big broadcasting organisation, to the city is fundamental in that respect. The media-city project looks promising as a means to make as much as possible out of MDR's relocation. A concentration and increase of production may also have a beneficial impact on film-making in the region, as film-makers and TV producers use much the same resources. Sufficient critical mass may induce artists and other film people to settle in the city, increase the supply of equipment and thus make the cluster more complete.

A sector with much development potential is software and new media. For this type of activity, the sophisticated electronic infrastructure in the Leipzig region is an asset, as it makes the city a good place for data- and information-intensive industries.

The media-industry is a people's business. Therefore, one of Leipzig's main strong points is its extensive and high-quality education and research facilities. Their yearly 'output' of skilled people in all media fields constitutes a source of talent and creativity with much potential. The challenge will be to keep media talent in the region.

If the opportunities and assets are numerous, so are difficulties, or even threats. In the field of TV and TV production, the cluster runs the risk of becoming overdependent on MDR as almost single commissioner. The nearness of Berlin is a serious restraint on the expansion of a film industry in Leipzig. Another weak point is the economic situation. The demand for new media and software products is relatively low in Leipzig, owing to its poor general economic conditions. Thus, the development of the local economic base is one of the preconditions for the multimedia and software sector to prosper, as these activities are strongly steered by demand from local firms. Finally, and following the same train of thought, too much young talent still seems to leave Leipzig for western Germany. As people are the key to any cluster development, this ongoing 'brain drain' needs continuous attention and can only be reversed by an integral approach.

References

Booz.Allen&Hamilton (1998), *Benchmarkstudie Electronische Diensten: 'Op weg naar de informatie-maatschappij', internationaal vergelijkend onderzoek t.b.v. de herijking van het actieprogramma Elektronische Snelwegen.*

Castells, M. (1996), *The Rise of the Network Society*, Blackwell, Cambridge.

Croteau, D. and W. Hoynes (1997), *Media/society: Industries, images and audiences*, Pine Forge Press, London.

Fidler, R. (1997), *Mediamorphosis: Understanding new media*, Pine Forge Press, London.

Freistaat Sachsen (1997), *Medienstandort Leipzig*, Sächsische Staatskanzlei.

Leipzig City Council (1996), Leipzig: Facts and figures.

McKinsey (1993), *Stimulating audiovisual production in the Netherlands*, Audiovisueel Platform.

Nederlands Economisch Instituut (1994), *De audio-visuele sector in Rotterdam: Economische betekenis, ontwikkelingsmogelijkheden en effecten Filmfonds.*

Rodekamp (1997), 'Messe im 20. Jahrhundert: Wendezeiten – Zeitbilder', in *Stadtgeschichtliches Museum Leipzig, Leipzig, Stadt der Wa(h)ren Wunder: 500 Jahre Reichsmesse Privileg*, Leipziger Messe Verlag.

Smits, B. (1998), *Rotterdam en de ontwikkeling van de AV-sector*, Audax Tros Multimedia.

Stadt Leipzig (1997), *Branchenhandbuch Medien Region Leipzig 1997/98*, Stadt Leipzig.

Stadt Leipzig (1998), *Leipzig Wirtschaftsbericht 1997/1998*, Stadt Leipzig.

Discussion Partners

Mr B. Bauer, Mitteldeutscher Rundfunk, Betriebsdirektion Grundsatzangelegenheiten, Head of Department.

Mr Th. Becher, Datafactory Informationssysteme, Managing Director.
Mrs G. Bock, City of Leizig, Mayor's Office, Head of International Relations Unit.
Mr A. Böswald, Leipziger Messe, Unternehmensprecher, Bereichsleiter Unternehmens-kommunikation.
Mr M. Friedrich, Saxony Telematics Development Corporation, Senior Project Manager.
Mr S. Grohmann, Sächsisches Institut für die Druckindustrie GmbH.
Mrs H. Gutsfeld, Logos Unternehmenskommunikation, Managing Director.
Mr G. Heyer, University of Leipzig, Department of Informatics.
Mr P. Hofmann, Sächsisches Institut für die Druckindustrie GmbH.
Mr M. Jähnig, Leipziger Gewerbehofgesellschaft mbH, Managing Director.
Mr U. Kromer, Leipziger Messe, Managing Director.
Mr B. Rittmeier, City of Leipzig, Geschäftsbereich des Oberbürgermeisters und des Stadtrates.
Mr C. Scheibler, City of Leipzig, department for Organisation and Services, Department Director.
Mr T. Weinert, Medien Pool.

Chapter Seven

The Health Cluster in Lyons[1]

1 Introduction

This case study is dedicated to the functioning and dynamics of the health cluster in the region of Lyons. Lyons has a long tradition in health care, medical research and education and hosts an impressive amount of firms and institutes related to health. Central to this analysis is the relationship between the actors in the health cluster. We identify private firms, hospitals, research institutes and educational centres. Furthermore, the development and potential of the cluster is put in the perspective of the spatial-economic structure of Lyons.

The organisation of this case study is as follows. In section 2, a brief introduction is given on the city of Lyons and its surrounding region, as well as its economic structure. Section 3 contains some general trends and developments in the health care and health-related industry. Section 4 focuses on the different elements of the health cluster in Lyons. The different actors and their activities are described: hospitals, firms, research institutes and education. After this descriptive section, in section 5 the focus turns to the dynamics of the cluster: an analysis is made of the interactions and linkages between the actors. In section 6, the functioning of the health cluster is put in the theoretical framework of cluster development in an urban context, as a means to put the cluster in perspective. Finally, in section 7 some conclusions are drawn.

2 The Economy of Lyons and its Region

2.1 Introduction

Lyons is situated in the southeastern part of France, in the region Rhône-Alpes, which is one of the 22 'Régions' of France. This region counts over five million inhabitants, and covers an area roughly the size of the Netherlands. Within the Rhône-Alpes region, there are eight 'départements', among which the département Rhône, to which the city of Lyons belongs. This département counts some 1.5 million inhabitants. Three-quarters of this population live in the Communauté Urbaine de Lyons. This area is often referred to as Grand Lyons, which is an urban region extending across the jurisdiction of four départements. Table 7.1 contains the key population data of the different administrative levels.

Table 7.1 Population in 1996 and development 1982–90

Unit	Number of municipalities	Population (x1000)	Development 1982–90
Région Rhône-Alpes	2,879	5,610	+6.6%
Departement Rhône	293	1,572	+4.2%
Aire Urbaine de Lyon	239	1,573	+6.7%
Communauté Urbaine de Lyon	55	1,309	+2.6%

Source: TEF, 1997.

Lyons lies at a short distance from the city of Geneva (150 km to the east) and the city of Grenoble (100 km to the southeast). It has a strategic logistic position: in the early 1980s, the city was connected to Paris by the first high-speed rail link in Europe. The TGV covers the distance to Paris in two hours Furthermore, Lyons has an international airport, Satolas (more than 5 million passengers in 1997).

2.2 Economic Structure

The economy of Greater Lyons is one of the most important in France. The region ranks second in France in terms of regional added value and employment. In addition, the economic growth in the Greater Lyons region is favourable compared to many other large French cities. Employment growth has been only moderate in the last few years, but still above the average for the larger French cities (see Table 7.2).

The economy of the greater Lyons region is very diversified. The region has a strong and modern industrial base, as well as a well-developed service sector. The service sector has grown rapidly during the last years, while industry and construction declined. See Table 7.3 for the economic structure of the Aire Métropolitaine de Lyon.

The production structure of Lyons is differentiated and comprises a relatively high proportion of knowledge-intensive activities. Within industry, the most important sectors in the region are chemical industry, pharmacy, equipment, automobiles, machines and metal industry. In Table 7.4, the most

Table 7.2 Evolution of total number of employees in 10 urban regions in France, between 1989 and 1996

	1989	1996	% change 1989–96
Paris	3,539,361	3,453,248	-2.43
Lyons	491,824	492,124	0.06
Marseille	321,151	321,614	0.14
Lille	307,033	307,418	0.13
Bordeaux	205,196	210,190	2.43
Toulouse	203,995	221,870	8.76
Nice	132,423	121,991	-7.88
Strasbourg	166,866	167,952	0.65
Grenoble	139,026	138,463	-0.40
St Etienne	89,833	84,841	-5.56
Average			-1.13

Table 7.3 Structure of the economy of Aire Urbaine de Lyon, 1996

Sector	Number of employees	Share, %	Abs. change 1989–96
Industry	133,918	27%	-21,960
Construction	36,700	8%	-8,204
Trade	87,076	18%	-724
Services	234,430	47%	+31,188
Total	492,124	100%	+300

Table 7.4 Employment in industrial branches, Air Urbaine de Lyon, 1997

Sector	Number of employees	Sector	Number of employees
Plastics	6,280	Extraction	320
Minerals	2,997	Food	9,382
Metallurgy	21,468	Textile/clothing	7,808
Machine building	21,372	Leather/shoes	727
Electronics	16,638	Wood	1,307
Automotive	11,841	Paper	7,092
Chemical industry	17,437	Nuclear power	1,286
Electricity/gas/water	2,093	Others	5,870

Source: Assedic, 1997.

important industrial sectors are depicted, in terms of turnover and employment. From the table, it can be seen that Lyons has a particularly strong position in machine building and chemical industry.

The service sector is also well developed in Lyons. Partly owing to its geographical position, Lyons has traditionally been a market centre, where various service activities are located, many of them with a supra-regional function. With some 60 banks located in the region, Lyons is also important as a centre of finance.

3 Context: Developments in Health and Health-related Activities

Before we turn to the description and analysis of the health cluster in Lyons, this section provides some trends and developments in health care and health related activities. Three developments will be focused on: the large increase in health expenditure; the rapid developments in biological and medical engineering; and the organisational developments in the pharmaceutical industry.

3.1 Health Expenditure Increase

One outstanding and well-known development in the health sector is the general increase of costs and expenditure on health. Figure 7.1 shows the upward tendency of expenditure on health care as a percentage of GDP for several countries. The figures are only relative: since GDP increased substantially between 1980 and 1996 in all the countries listed, the absolute expenditure would be even more impressive. There are several reasons for the increase in health expenditure. In the first place, demand for health services is growing because of the ageing of the population in many Western countries, implying a increase in chronic diseases and disability. Secondly, the pace of technology development has quickened over the past decade, generating new diagnostic techniques and the treatments for conditions for which there was previously no treatment (World Health Organisation, 1997). There is strong pressure on health care providers to adopt the latest available medical techniques and treatments and a strong incentive for pharmaceutical industry to develop new products.

The Western countries differ greatly in the way their health systems are organised. In the USA, the health system is based on market principles; individual citizens have their own responsibility and are expected to take care

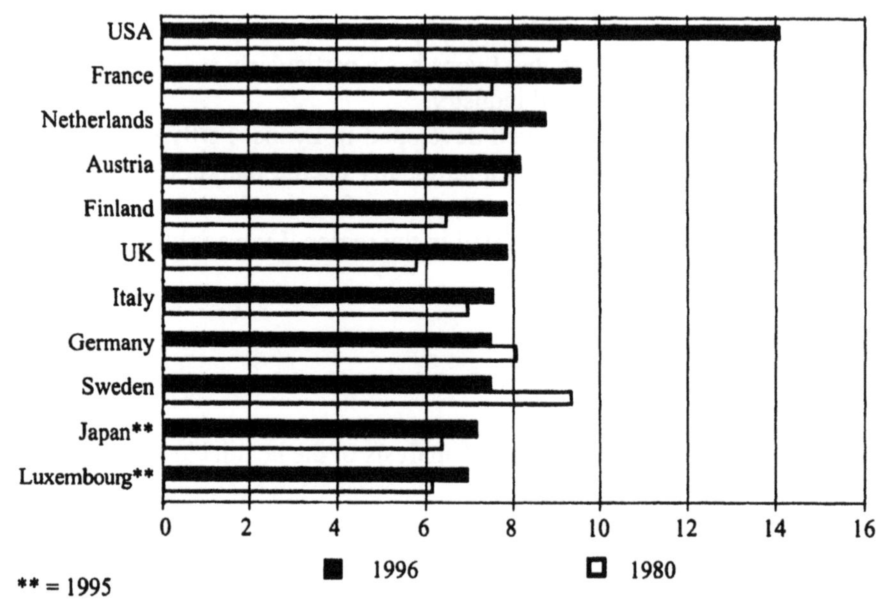

Figure 7.1 Health expenditure as % of GDP in different countries, 1980 and 1996

Source: Tableau de l'Économie Française 1998–99.

of their own insurance. Government intervention is limited to control mechanisms and quality regulations. Exceptions are made for the poor, who can make use of the state-financed medicaid programme, and the elderly, for whom there is the Medicare programme. A substantial part of the American population, some 15 per cent, have no insurance. In Western Europe, the role of government in the health care provision is much greater. In France, medical aid is financed through the social security agencies. Many French have additional private insurance provisions. The United Kingdom has the National Health Service system, financed by taxes and open to everyone. The differences in government intervention in different countries can be seen in Table 7.5, which depicts the public expenditure as a percentage of total health expenditure.

Under pressure of steeply rising health costs and the simultaneous need to cut government budgets, Western countries are trying to limit the growth of medical expenses. They pursue different strategies to achieve this. In the UK and Germany, efforts are made to save expenditure by introducing market elements into the health sector, such as competition amongst insurance companies. Austria tries to save on health costs by, among other things, introducing performance-oriented hospital financing, by obliging social

Table 7.5 Public part of health expenditures in %, 1994

Country	% public contribution
Germany	74
United States	45
United Kingdom	84
France*	75

* = 1997.

Source: http://www.sante.gouv.fr/sesi/kebec/syst-san.htm; Tableau de l'Économie Française.

insurances to finance preventive medicine and promote health, and by reducing financial support in some cases (Federal Ministry of Health and Consumer Protection, 1996). In France, the national government put quantitative restrictions on most categories of health services to reduce the growth of expenditure at the beginning of the 1990s. Since November 1995, new reforms have been introduced. The keyword is regionalisation: regional agencies of admission have been established, as well as regional medicaid agencies.

3.2 Biological and Medical Engineering

As stated in the last section, one of the causes of increasing costs in health is rapid technological development. Progress is particularly fast in microbiology (which forms the basis of the development of new drugs and substances). The progress is fuelled by increased R&D expenditure – both of firms and of governments – and new scientific discoveries. Another factor is the increased productivity in research and development. In particular, the potential of information technology and robotics has entailed new methods to develop new substances more systematically and rapidly. The rapid change in biological and medical engineering is reflected in a turnover growth in this segment of 400 per cent from 1980 to 1996 (ERAI, 1998). The world market of biological and medical engineering now amounts to one quarter of the total pharmaceutical market.

US firms clearly dominate the scene: 50 per cent of world production takes place in the USA, 27 per cent in Europe and 20 per cent in Japan. Table 7.6 shows the dominance of the USA vis-à-vis Europe in the biotech industry. Although the number of firms is only slightly higher in the USA, the lead in research and development is enormous.

The USA's lead over Europe, in particular in the field of gene technology – the fastest growing segment within biotechnology – can be explained by

Table 7.6 The biotech industry in USA and Europe 1997, some indicators

	USA	Europe
Number of firms	1,274	1,036
Total employment	140,000	39,045
R&D expenditure (in m. Ecu)	8,268	1,910

Source: Ernst and Young, 1998.

several factors. First, the USA federal government supported fundamental research in this field from as early as World War II. Second, since the second half of the 1970s, relations between universities and business have improved rapidly in the USA, resulting in more efficient and effective research and high levels of new firm creation (OBIG, 1998). Consequently, the number of patents is considerably higher in the USA than in Europe. Between 1992 and 1994, the USA registered 49.5 per cent of all patents in biotechnology. Europe, at some 33 per cent, lags behind considerably (Clement et al., 1998).

3.3 Organisational Changes in the Industry

The competition between companies in the pharmaceutical and biotechnological industry is severe. As the future market position of firms is considered to depend more and more on biotechnological innovations and the product life cycle is shortening, research and development is of strategic interest. Different strategies are pursued to speed up innovation by increasing flexibility and creating scale. On the one hand, established companies seek small companies with innovative products to build their diversification for the future. On the other, to generate scale economies in research and marketing, mergers, acquisitions, joint ventures and strategic alliances are the order of the day (Arthur Andersen, 1996). For instance, at the time of writing, the German pharmaceutical giant Hoechst is about to engage in a strategic alliance with Rhône-Poulenc, a French pharma-group. Together, they aim to become a world leader in life sciences. The main object for cooperation is to create a degree of scale large enough to allow them to share the enormous research and development costs (*Financial Times*, 25 November 1998).

The world market is now dominated by a small number of players, the most important ones being from the United States, Germany, Switzerland and UK. The restructuring wave, which is not over yet, has (and will have) consequences for the location of the firms' various activities (R&D, production, logistics). The keyword is rationalisation, implying a decrease in the number of locations on a world scale. For instance, Hoechst plans to concentrate all of its R&D activities in New Jersey, USA, implying a loss of R&D activities in several German cities (NRC Handelsblad, 26 November 1998). Additionally, most groups do not invest in new sites but concentrate research or production at the location of one of the merging or cooperating partners.

4 Actors in the Health Cluster in Lyons

In this section, the main actors in the health cluster in Lyons are described. A distinction can be made between the following types of actors: private firms, hospitals, research institutes and education institutes.

4.1 Firms

It is not easy to make a clear distinction between firms in the health cluster on the basis of their activities. In the most straightforward distinction, the cluster in Lyons could be divided into five categories: the pharmaceutical industry, biomedical technology, the veterinary industry, paramedical firms (homeopathy) and medical instruments. However, the large pharmaceutical firms often have a biotechnology department for the research and development of new products. The same holds for the veterinary and the paramedical firms. Therefore, the biotechnology firms will not be treated separately, although there certainly are firms who specialise in this field.

Pharmaceutical industry/biotechnology In this branch, the Rhône-Alpes Region holds a prominent position in France. The production of the most important parts of this sector – medicines and in vitro diagnostics – amount to 15 per cent and 30 per cent of the French national production respectively. For this sector, the presence in the region of complementary activities such as chemical industry and medical textiles is a clear advantage. In the field of biotechnology, the position of Lyons is particularly strong because of the presence of components of Mérieux, a dominant firm in the region active in several branches of biotechnology. The most important firms are depicted in Table 7.7.

As can be seen in the table, many of the firms belong to international pharmaceutical conglomerates. In part, the internationalisation of the Lyons industry has been a result of recent mergers, acquisitions, joint ventures and

Table 7.7 Large pharmaceutical firms in Lyons

Firm	Organisation	Activities/products	People employed
Lab Marcel Mérieux	Independent	Clinical biology; molecular biology and genetics	Lyons: 340
Pasteur Mérieux Connaught	100% Groupe Rhone Poulenc	Production and development of human vaccines/serums	Worldwide: 5,600 France: 2,547 mostly in Lyons
Pasteur Merieux MSD	50% Pasteur Mérieux Connaught, 50% MSD (Merck Sharp & Dohme)	European distribution of products of Merial and Pasteur Merieux Connaught	Worldwide: 750 France: 250, mostly in Lyons
Bio Merieux	Independent	In vitro diagnostics, in particular infection diseases	Worldwide: 3,300 France: 1,730
Chiron-Domilens	Joint venture with Chiron, San Francisco	Eye prostheses	Worldwide: 6,500 France: 200
Aguettant	Independent	'Solutés injectables'	n.k.
Lipha-Merck	Subsidiary of Merck, Germany	Diabetology	Worldwide: 2,900 France: n.k

Source: Aderly, 1998.

strategic alliances. Even a large player like Mérieux has therefore had to cooperate with large firms from abroad. In general, the Lyons region does not seem to have benefited from this rationalisation (Arthur Anderson, 1996). In many cases, the headquarters of the conglomerates are not located in Lyons, but elsewhere. For example, Lipha Merc is directed from Darmstad, Germany, and Pasteur Mérieux MSD from London. This increases the external dependency of the health cluster in Lyons. Although several firms mentioned above have laboratories in Lyons, which binds them more ore less to the region, the research projects are often programmed abroad. One example is the research programme of Schering Plough in Lyons, which is decided upon in the firm's headquarters in Philadelphia, USA.

Paramedical industry and medical equipment/ instruments In the paramedical subsector, one firm is clearly dominant in Lyons: Boiron, world leader in homeopathic products. This firm has its headquarters in Lyons, as well as several R&D and production facilities. It is firmly rooted in the region, mainly because the Boiron family, strongly tied to Lyons, still plays a major role in the firm. The company finances its fundamental research – especially on the working of homeophatics – through foundations. Boiron expects more fundamental research to be conducted in the future.

A second important subsector is medical equipment. The sector in the Rhône-Alpes region produces 20 per cent of the national output, most of which is in Lyons. This sector is very diversified. The products range from surgery instruments to wheelchairs and from artificial organs to medical textiles. The firms in this sector are generally small (see Table 7.8), although there are exceptions, such as Dermscan and Radiometer Analytical.

Table 7.8 Size distribution of producers of medical devices, Rhône-Alpes region

Class (number of employees)	% of firms
>500	3
100–499	13
50–99	17
20–49	23
<20	44

Source: Arteb, presentation.

Veterinary industry A particularly strong point for Lyons is the veterinary industry. The region of Rhône-Alpes provides 68 per cent of the veterinary products of France. This position is based on, among other things, the role of Lyons as centre of the agricultural region of southeast France. A few large firms dominate the scene. One of them is Mérial, a 50 per cent joint venture with Rhône Poulenc and Merck. The firm is located in the Gerland area. It works closely together with a technology centre of Rhône Poulenc dedicated to vaccines and protein separation. It is the largest producer of veterinary products in the world. The firm shares knowledge with Pasteur Mérieux Connaught, which is active in human veterinary products. Mérial is eager to discover new technologies that might have commercial potential. For that purpose, the firm has at its disposal networks of experts all over the world who scan both other firms and universities for potential expertise. The role of Mérial is then to translate the licence into a saleable product.

There are a number of smaller, specialised firms and institutions in this region, which have a high growth potential. Examples are the Institut de Phytotherapie Animale, a high tech institute in which dietetical and hygienic animal products are developed, and IFFA-Credo, a subsidiary of Charles Rivers Group, USA, specialising in the production of laboratory animals.

Private laboratories Besides research as a part of manufacturing activities, the region also hosts several independent private laboratories that work for either private or public principals. Examples of public principals are the World Health Organisation (WHO) and the research programmes of the European Commission. Given their functioning in an international competitive setting, the private labs in Lyons have to work according to the highest standards. Consequently, their efficiency and quality are high. Most labs employ an international staff.

The state-of-the-art laboratory for virological research named 'laboratoire P4 Jean Mérieux' – officially opened in December 1998 – is an example of the level of private facilities. This laboratory is unique in the world and has been financed by the Jean Mérieux Foundation in Lyons. The laboratory is located in the Gerland business park near the Merial establishment. It is expected to function as a growth pole for related activities in research and advice.

4.2 Hospitals

Lyons has 18 hospitals, offering 6,505 beds. This makes it the second hospital concentration after Paris. The hospitals employ 22,000 people, thereby being by far the largest employer in the region. They have a budget of FF 6.5 billion a year. Most of the hospitals are general purpose, but some are specialised. For example, there is a large neurological hospital, with 429 beds, and a hospital specialising in heart disorders. Besides variety in function, each hospital has its own character. The pluriformity of the hospitals is a strong point of the Lyons hospital system.

The hospitals have a local, regional, national and sometimes even international function. For a long time Lyons was the only reliable and high quality hospital centre for the South of France. Wealthy people came to Lyons for treatment. Although that regional importance has declined with the arrival of high quality care in southern cities such as Marseilles, Toulouse and Grenoble, the specialised hospitals in particular still have a supra-regional function. For example, many Italians are treated in Lyons, especially in the care fields of neurology, cardiology and urology. This is mainly so because these disciplines are very poorly developed in Italy, although recently much progress has been made.

The management of the hospitals is centrally organised. The 18 hospitals are directed by a central organisation called Hospices Civils de Lyons (HCL). This organisation owns the land and is responsible for the management of these hospitals. The HCL is not an independent regional organisation: it is controlled directly by the central government of France.

The Lyons region counts several private clinics, although increasingly they are being integrated with the public health care system. Stimulated by health insurance organisations seeking higher efficiency in the health care system, hospital functions are coordinated and clinics are merged to realise scale economies. However, with the integration of private clinics, their unique character – the 'personal touch' – is at stake, which will make the Lyons hospital system more uniform.

The hospitals do not only have a function in health care: they are also important locations for the research activities of the medical faculty of the university. One half of university medical research takes place in the hospitals. Furthermore, the hospitals host 12 research institutes financed by INSERM, a national organisation that finances health research. The fields of research in the hospitals are selected and approved by this organisation. This does not mean that the hospitals have no influence on the selection of research themes: they recruit skilled and successful scientists in the fields they want to concentrate on, who are in turn financed by INSERM. A further important research fund is the Programme Hospitalier de Recherche Clinique (a programme for clinical research in hospitals). This programme has a budget of FF 6 million a year, and is mainly directed toward research into medical instruments and equipment.

Location developments In 1988, the HCL and the city administration decided to reduce the number of hospitals and to concentrate them at three locations in the city. The mayor initiated the decision. Since then, three concentration poles have been created, in the north, east and southern part of the city respectively. Each pole contains a concentration of health care, education and research activities.

The main reasons for the concentration of hospitals at three locations were the needs both to reduce costs and to create synergies by combining medical specialisations. Cost reductions are gained by sharing very expensive facilities and instruments. In addition, spatial concentration implies considerable savings on 24-hour services compared to a situation with hospitals dispersed all over the city. Another expected benefit of physical clustering was the possibility to 'break down the walls' between medical disciplines. For example, at the east pole a new building has been constructed in which neuroscience and paediatry, two disciplines that used to have their own establishments, have been combined. This should result in new combinations and fruitful synergies. A second example at the east pole is the combination of obstetrics and neonatology in one location. Here also, as well as cost reductions, important synergies are expected in terms of research and care.

Although there is no clear specialisation amongst the poles, they differ in many respects. The north pole is the least important of the three. Here, the research and education functions are virtually absent and the number of beds is considerably smaller than in the other two poles. There is some specialisation in AIDS, hepatitis treatment and toxicology. The south pole specialises in the fields of cancer, diabetology and nutrition. It counts three large hospitals. The east pole is the largest and most important pole, both in terms of scale and specialisation. The large Heriot hospital, with 1,235 beds, is located in this pole. The pole hosts hospitals specialising in cardiology, paediatry, gynaecology and neurology. Within the east pole, a strong neurological focus seems to have emerged, with a specialised neurological hospital and high quality research activities in the field of neuroscience. There are research groups from the university and INSERM (a French organisation responsible for financing medical research). Recently, a special institute for cogni-science, a branch of neuroscience, was opened at the east pole. This national institute is financed by INSERM and is an important addition to the development of the east pole. Its location in Lyons has been stimulated by the Region Rhône-Alpes, which bought the land for establishing the institute. The institute is developing nonexclusive relationships with the Lyons universities and hospitals.

The total cost of the spatial reconfiguration of hospitals in Lyons amounts to FF 5 billion, the greatest part of which is financed by the HCL. The city of Lyons contributes FF 200 million and the Communauté Urbaine 300 million. Of this amount, FF 40 million is invested in infrastructure and public facilities. The operation should be ready by 2004.

4.3 Public Research Activities

Public research in Lyons takes place on a large scale: the city hosts an important concentration of universities and research institutes. The most important specialisations in the research in he region are developmental biology (virology, genetics), neurology, and various kinds of cancer research.

The most important university for the health cluster in Lyons is Claude Bernard University. Many research institutes in the region are attached or related to this university. The university research institutes financed by national research organisations are very important. The two main organisations are CNRS (Centre National de la Recherche Scientifique, for science in general) and INSERM (Insititut National de la Santé et de la Recherche Médicale, for medical research). Some of these public research institutes and laboratories undertake contract research for the industry.

The strategy of the Claude Bernard University is to concentrate research activities in a few important fields. These fields are neuroscience, cardiovascular research, endocrinology and immunology/virology. Each field consists of research themes with further specialisation. The specialisation strategy is supported by INSERM and CNRS. However, coordination of the directions of specialisation with private sector research wishes is very limited.

Apart from the Claude Bernard University, an important role in university research is played by the École Normale Supérieure. It is a prominent university to which only excellent students are admitted. The research activities of this university are particularly strong in genetics and virology. In general, the laboratories are not very large. A third important institute is the School of Engineers related to INSA (a French institute for applied science). This school counts 3,000 students and carries out applied research in the fields of engineering and life sciences.

Health-related research by public institutions in Lyons is entirely financed by public means. Contract research is conducted only on a very limited scale. Given the general budget reductions for universities, public research centres increasingly have to accept third party financing. However, many groups in the Lyons public research structure do not seem to be well equipped to conduct research for the market. They lack professionalism in terms of efficiency and quality; scientists often are unaware of international research protocols and existing patents. Consequently, their research output is of little use in the market.

Locational developments The research activities related to the health cluster are conducted at different locations. Firstly, research takes place at the faculties of the university. Second, much research is executed at the hospital poles described in section 4.2. This mainly concerns clinical and experimental research activities. Although all three poles contain research labs, the east pole is by far the most important location for this type of research, mainly because of the specialised hospitals that are located here. A third important research area is the Gerland area. This prestigious area hosts an interesting mix of firms (of which Mérieux is the most important; others are Aguettant and Domilens), private and public research institutes, and the École Normale Supérieure.

In due time, a part of the Claude Bernard University is to be moved to the research park of Gerland as well. Very importantly, the university will open new labs and other research facilities in Gerland. An important research lab in the Gerland area is ICBP (Institute of Chemistry and Biology of Proteins), specialising in protein research, with 120 researchers. It is linked to the university and largely supported by the CNRS.

Fourthly, much research (both public and private) is carried out at the campus called La DOUA. The science faculty of the university is located here, with a total of 210 labs, with 145 life sciences labs and 12 biological labs. Next, there is the before-mentioned INSA (Institut National de Sciences Appliquees de Lyon). This institute is directed towards applied sciences. Of a total of 31 research labs, the institute counts six life science labs. In total, this La DOUA campus counts some 3,000 researchers and 20,000 students. Apart from this, there are private firms established on the campus, some of them specialising in pharmacology or biotechnology.

4.4 Education

Lyons is an important educational centre: the city counts some 100,000 students in higher education. It is the second most important university concentration in France, after Paris. The division of the total number of students in higher education is depicted in Table 7.9.

Table 7.9 Students in Lyons

Discipline	Percentage
Sciences and medicine	28%
Letters, social sciences/human sciences	45%
Commercial and management	2%
Engineering	8%
University preparation	12%
Others	5%

Source: Aderly, 1996.

The most important university is the Claude Bernard University. It is divided into three parts: sciences and medicine (27,036 students, mostly in the DOUA campus); social sciences (24,621 students); and literature (18,610 students). For the health cluster, sciences and medicine are the most important sections of the university. Within medicine, the division of students is as shown in Table 7.10.

There is also a special veterinary school with over 500 students. Apart from education, fundamental and applied research in the fields of animal virology, pathology and nutrition is carried out at this school.

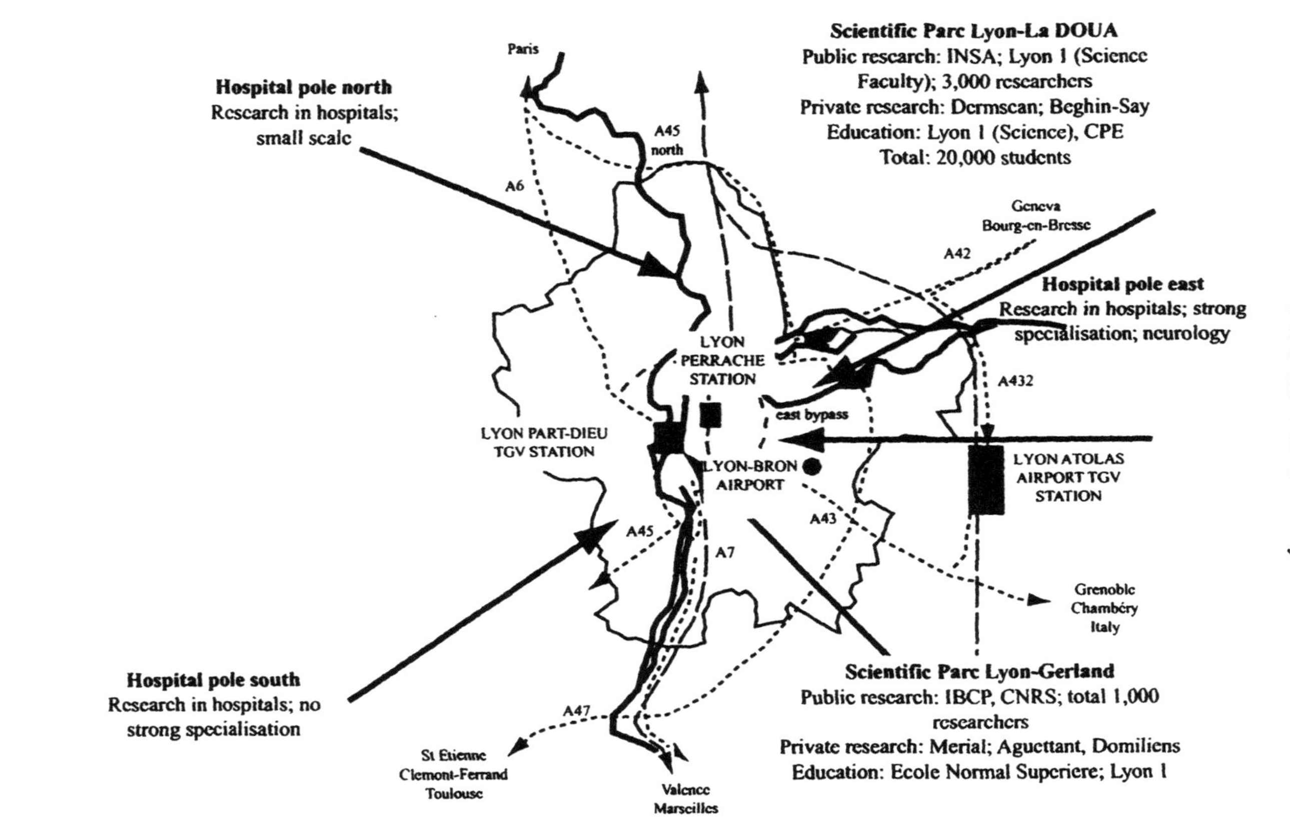

Figure 7.2 Locations of medical research activities

Source: Aderly, 1996, adapted.

Table 7.10 Medical students

Section	Number of students
Medicine	6,799
Pharmacy	2,061
Dentistry	487

Source: Aderly, 1995.

A unique and very important element in the health cluster is the École Normale Supérieure. This university, created in 1987, is accessible only for very good students. Apart from in Paris, there are no other establishments of this school in France. This university counts over 500 students, many of them specialising in life sciences. The École Normale Supérieure has at its disposal or participates in 12 labs, among which are labs for molecular and cellular interaction, chemistry and biochemistry, virology and molecular biology. The students of this university are very internationally oriented: about a quarter take apprenticeships in a foreign country. The École is situated at the Technopole of Gerland, amidst large biotechnological firms and other research institutes. Higher education is also related to hospitals in Lyons, where many students follow practical courses.

In the medical and biological fields, Lyons counts no specialised courses or postgraduate education that is sponsored by private business, although there are some initiatives. For example, Mérial, one of the large veterinary producers in Lyons, has had plans to set up a education programme at the university, together with Mérieux. However, because of day-to-day turbulence in the business, this project has not yet been realised, although the management is clearly convinced of its importance.

4.5 Local and Regional Policies

As a final actor in the cluster we will discuss the role of local and regional government in the health cluster in Lyons. In the first place, the Communauté Urbaine, consisting of Lyons and surrounding communities, has recently drawn up an action plan for the future of Lyons as a technology centre. In total, FF 21.5 million is available. Special attention is paid in this to the health sector, which is clearly regarded as a centre of excellence for the region. Specific actions in the health cluster are:

- the stimulation of creation of new firms;

- the improvement of Lyons's image in the field of life sciences;
- the development of a strategy for the cluster as a whole.

In this strategy-building process, all actors in the cluster are involved, as well as ADERLY (the department of the Communauté responsible for the attraction of firms from elsewhere), the region Rhône-Alpes and the state. The Grand Lyon region contributed FF 3 million in total, over a period of three years.

A second important element of the policy of the Communauté is the further development of the Gerland area. In particular, Gerland is to become a physical focal point of research, business and education in the biomedical and biotechnological field. The Communauté is especially keen on stimulating cooperation and network building in the area.

Another spatial concentration point is the location of east pole Lyon Bron. The Communauté participates in the construction of a new specialised hospital and takes action to further cooperation and stimulate synergies (Communauté Urbaine, 1998a).

Finally, Biovision can be mentioned as an element on which to build the health sector in Lyons and give the cluster international status. Biovision is an international biannual conference and trade fair, held for the first time in March 1999. Biovision aims to make Lyons the 'Davos' of the international health care industry (the Swiss city of Davos hosts the world forum on international economics affairs). The objective is to examine the current situation of life sciences and its future challenges. The conference is to provide a framework for discussion among industry, government, research and consumers, necessary for the development and acceptance of new technologies. The initiative for Biovision has been taken by Lyons' mayor Raymond Barre together with key people from the industry and the research institutes. For Lyons, the manifestation is a city-marketing instrument to put the region on the mental map as an international centre of health care.

5 Dynamics in the Cluster: Interaction and Spin-offs

As described above, Lyons hosts an impressive body of health care institutes both specialised and general, many firms, some of which very innovative and internationally oriented, high quality research activities related to the medical activity, and many students in science and medicine. In this section, we take a step further, by analysing the dynamics of the cluster. The relations of the different actors in the cluster are described, as well as linkages to actors outside the cluster (see Figure 7.3). The numbers in the figure correspond to the numbers of the subsections.

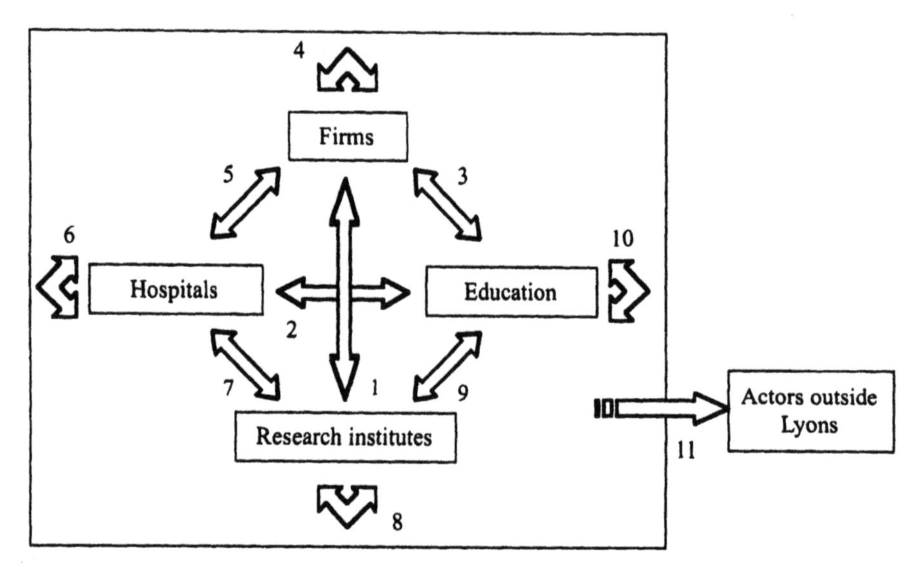

Figure 7.3 Interactions in the cluster

5.1 Interaction between the Business Sector and the Research Institutes

Generally speaking, the relations between business and the public research infrastructure are poorly developed. Too often, the research labs and firms operate completely separately. There are many reasons for this. In the first place, applied and contract research are not highly esteemed by scientists. The prevailing view in the scientific community is that science should be pure. Practical applications are suspect. In the second place, and related to this, the status of scientists is determined to a large extent by the number and quality of their scientific publications, not by practical results. This makes applied research less popular. In the third place, the national research agencies (CNRS and INSERM are the most important) determine to a large extent which research is financed and which is not. As a result, the research themes are often not well adapted to business needs. It is difficult to finance research through contracts with business, although the situation is changing slowly but steadily: CNRS, responsible for the repartition of research funds, in some cases even sets the condition that research is to be partly financed by non-public funds. In the fourth place, there are large cultural differences between researchers and private firms. This is a very important barrier to interaction.

Even though the difficulties are serious, there are some good examples of the successful combination of research and business. This holds particularly in the field of biology and biotechnology. Most of the cooperations are on a small scale, but they mark a turning point. At the scientific pole of Gerland, where there is a concentration of firms and research institutes, ICBP laboratories (Institute of Chemistry and Biology of Proteins) participate in projects together with industry, mostly Mérieux. The cooperation takes three forms: first, the firm directly funds research programmes in the lab; second, it finances students, PhDs and researchers who work in the lab on a project that is of interest for the firm; and third, they pay for particular analyses. More in general, the close physical proximity plays a role in interaction in Gerland. In particular, the fact that people from the labs, the firms and the École Normale Supérieure take lunch in the same restaurant is important, as it facilitates contact and stimulates cooperation.

There are some official organisations that are aimed at facilitating or stimulating the linkages between the research institutes and universities on the one hand and the business sector on the other. For example, there are organisations that aim at linking business with universities. One of them is EZUS. This independent organisation was initiated by the Claude Bernard University in order to execute financial and managerial handling of projects in which industry and university cooperate. EZUS figures show that the amount of contract research is rising steadily: see Table 7.11. As becomes clear from Table 7.12, by far the largest part of contract research is done for the private sector, in which the pharmaceutical industry plays a considerable role. It is not only the Claude Bernard University that has a contracting organisation: a similar structure exists for the INSA, called INSA-valor, founded 10 years ago. In our interviews it was indicated that both EZUS and INSA-valor are handicapped by a lack of in-depth knowledge in attractive fields of research and the quality of research groups within their institutes. Consequently, they are not effective as brokers between private sector demands and the supply at universities and other public research centres.

Besides the transfer agencies of INSA and the Claude Bernard University, there are organisations that are responsible for the active promotion of industry and research cooperation. The most important is the non-profit foundation ARTEB, founded three years ago by the region Rhône-Alpes. Their mission is to find out what the industry needs in terms of research, to build a network of competence in the region, to stimulate cooperation in research and development, to help firms with regulations and to help small and medium sized enterprises (SMEs) in internationalising. ARTEB focuses on the biomedical industry and

Table 7.11 Number and value of contract activities intermediated by EZUS, 1990–97

Year	Number of operations	Value (in million FF)
1990–91	131	17.4
1991–92	240	17.7
1992–93	306	24.2
1993–94	426	29.8
1994–95	540	37.7
1995–96	791	51.9
1996–97	996	65.5

Source: EZUS, 1997.

Table 7.12 Distribution of contract research in 1995–96

Type	Number of operations	Value (million FF)	Percentage
Public and semi-public organisations	76	2.5	5.7%
Large firms	188	17.0	38.7%
Pharmaceutical labs	132	11.1	25.4%
Small and medium sized enterprises	302	10.8	24.5%
Others	42	2.4	5.7%
Total			100.0%

Source: EZUS, 1997.

research, mainly because of the relatively small size of firms in that sector.

For the pharmaceutical industry, which consists mainly of large firms, technology transfer is less important. These giants carry out their own research or buy it elsewhere. Still, there is a very small organisation named AFIPRAL (one employee only), that tries to stimulate networks between firms and labs in the region.

5.2 Interaction between Education and Hospitals

There are many interactions between hospitals and educational institutes in the health cluster. As stated in section 4, the hospitals fulfil an important function in the education of medical students, who get practical clinical and research training at the hospitals in Lyons. This interaction is fruitful: students are trained in an environment similar to the one they will work in when they have finished their specialisation, and hospitals have an additional research capacity.

5.3 Interaction between Education and Firms

There are many different types of relationship between firms in the health cluster and education. Firstly, many students from education institutes get on-the-job training in firms. This holds in particular for students in applied science and engineering, and less for medical students. Secondly, some firms, mostly large ones, employ or finance PhD students or graduates to execute specific research. This does not, however, occur one a large scale.

Thirdly and very importantly, educational institutes can function as breeding ground for young entrepreneurs who want to start a new business. The number of starters is very restricted, however. Only a few dozen new medical or medical-related firms were created in the last 20 years. This number is small, in particular compared to firms' birthrates in the US. There are several reasons for this:

1 there is a lack of entrepreneurial attitude among students. This is a problem in France in general. In the case of the health cluster in Lyons, the attitude of teachers plays a role. Very often they promote a negative attitude about business (businesses are not scientific, only aimed at money making, etc.), largely because they do not know much about business. In that respect, the lack of commercial education of biologic/medical students is also important;
2 it is very difficult to start a medical firm (biotechnology or instruments) because of rigorous and often unclear regulations. There is no organisation that helps young firms to overcome this barrier;
3 venture capital is scarce. Unlike in the US, banks are generally not willing and prepared to invest money in risky and uncertain projects. This often means that young starters are dependent on the informal help of friends and family to start their business;
4 accommodation and equipment often poses a problem for new businesses. Equipment is very expensive, particularly for medical and biotechnology. There are no structural programmes to help firms to find accommodation or purchase equipment in the first few years.

Very importantly, young firms are often firmly rooted in the region: they are dependent on a network of friends and former teachers. This embeddedness of small firms is important to reduce the external dependencies of the cluster, particularly at the present time, when the existing large firms are increasingly directed and steered from outside Lyons.

5.4 Links between Hospitals and Firms

Although hospitals often buy their medical equipment from medical firms in the region, the number of strategic linkages between hospitals and firms in the regions is limited.

It is illustrative that at the three concentration poles of the hospitals (east, south and west), there is no space for private firms. Only care, research and education are allowed. Even so, there are examples of strategic cooperation between hospitals and firms. One example is the joint development of a machine to treat prostate cancer by a hospital and a firm called Technomed. They have combined their knowledge of the treatment of cancer and the building of machines respectively.

5.5 Interaction amongst Hospitals

In the recent past, the many different hospitals in Lyons hardly cooperated in any field. The lack of cooperation entailed high costs, as high-grade facilities had to be present in every hospital. In addition, functional synergies were lost because of the limited interaction between hospitals. However, as stated in section 4.2, with the physical concentration of hospitals in the three poles, the situation is changing drastically. Examples of recently-established cooperations are the establishment of a new building in which neuroscience and paediatry are combined, and the sharing of facilities and 24-hour services in the east pole. Because of their cost-saving and innovation-stimulating effects, the increasing inter-hospital interactions have a beneficial impact on the hospital cluster in Lyons as a whole.

5.6 Interaction between Hospitals and Research Institutes

The link between hospitals and clinical research – most of which takes place within the hospitals – is intense. Apart from the 'normal' clinical research, some more specialised hospitals have dedicated research institutes financed by INSERM of CNRS. The policy of physical concentration and functional cooperation of hospitals has a beneficial impact on research in the hospitals, in particular on the more specialised research. This is particularly true of the east pole, where a strong concentration of specialised research in the field of neuroscience is being developed.

5.7 Interaction amongst Public Research Institutes

Amongst public research institutes, interaction often falls short. For instance, there is virtually no interaction between the medical faculty of the university and the International Research Centre for Cancer, which is a part of the World Health Organisation. This centre executes much research on cancer incidence and gathers worldwide information on the subject. The absence of interaction with local researchers is all the more surprising as the centre receives scientists from all over the world.

The recent initiative of the central organisation of hospitals to construct new buildings in which specialisations are combined, in order to break down barriers between the formerly separated specialisations, is promising for the improvement of research cooperation in the health sector. In the east pole in particular, interesting research combinations are blooming.

A second field in which research combinations could be fruitful is in the cooperation between the science department and the medical department of the university. Despite many opportunities for synergies, the linkages in research are still scarce. One good but single example is in physiology, where people from science (engineers) work together with neurology experts on specialised equipment for cogni-science, a specific branch of neuroscience.

5.8 Interaction between Research Institutes and Education

There are many interactions between educational institutes and research institutes. Very often, research and education are in the hands of the same institute, which facilitates interaction. For one thing, teachers or professors at the university are simultaneously active as researchers in a laboratory. For another, many university departments (in particular in the biomedical field) have research laboratories in which students work on their theses.

Cooperation is not limited to the public research infrastructure: private research institutes as well sometimes employ regular or PhD students to execute specific research projects. For instance, Schering Plough offers a dozen of PhD positions each year.

In some cases, cooperation between research and education proves difficult. An example is the relation already mentioned between the medical faculty of the university and the cancer research centre of the World Health Organisation. Although the Cancer Institute could very well offer educational courses in the fields of epidemology and public health, two fields in which the institute is very strong, linkages are almost absent.

5.9 Interaction amongst Education Institutes

The interaction amongst education institutes is rather poor. The low level of interaction is at the expense of the volume and quality of interdisciplinary research in Lyons and the possibilities to generate scale economies in undertaking research and exploiting facilities.

5.10 Relations outside the Cluster

Participation in international networks is essential for the long term success of any research institute. This is particularly true in medical and bio-technological research, where developments are fast and on a world scale. The large biomedical firms in Lyons recognise this, and operate internationally. The large producer of veterinary products Mérial, for instance, actively exploits a network of experts all over the world, who are charged with the scanning of the latest technological developments that might be of interest to the firm. If Mérial thinks a technology is promising, it buys the licence and turns it into a commercial product. In general, the large firms outsource much research to top institutes abroad.

Smaller firms and universities in Lyons participate much less in international networks. The relative closeness of the health sector is a barrier, as is the lack of English speakers. There are positive exceptions. One is the ICBP, which actively participates in international and European research programmes. The same holds for some research groups at INSA (school of applied science).

6 Confrontation with the Framework

In this section, the health cluster in Lyons is placed in the perspective of the city as a whole (urban economic structure, spatial and cultural conditions), and the organising capacity of the cluster is analysed.

6.1 Spatial and Economic Context of the Cluster

The *economy* of Lyons is well developed and on the whole strong, in particular compared to the rest of France. The economic structure of the region is diversified, in terms of the division between sectors and size classes of firms. For the health cluster in Lyons, the presence of a strong logistics sector and of the

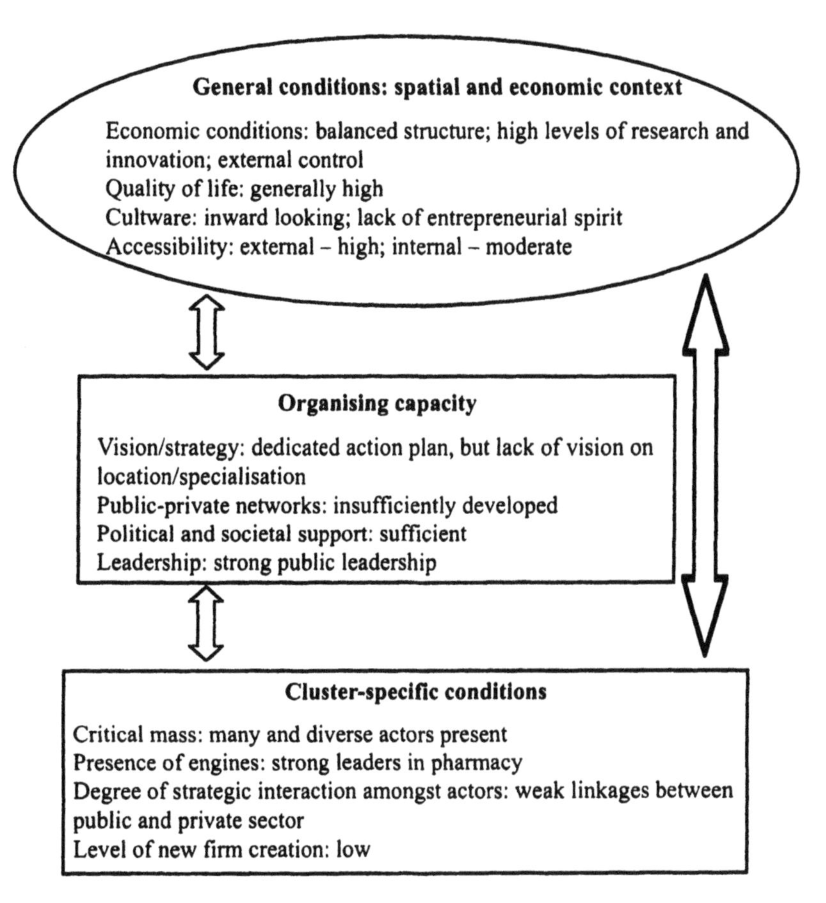

Figure 7.4 Framework of reference

chemical industry especially offer clear advantages. A weak point for the region of Lyons (which also applies to the Rhône-Alpes region) is the small number of corporate headquarters. This makes the region relatively vulnerable to external changes. The knowledge base of the Lyons region is well developed. The skills of the population are generally high and the region counts a large concentration of higher education institutes. There are also relatively many research and development activities in the Lyons region. This creates a favourable 'innovative milieu'. However, as has been made clear in the previous sections, there is a bottleneck respecting the practical application of scientific knowledge. This is due, among other things, to a lack of 'professional skills' and interaction between public research and education on the one side and (international) private research and manufacturing on the other.

The '*cultware*', being the soft part of the spatial economic conditions, shows some weak points. In the first place, Lyons shares the problem of a lack of entrepreneurial spirit with the whole of France. This holds not only for the attitude of the universities and research institutes, but also for teachers and students. It reduces the economic dynamism of the region. In the second place, most of the population of Lyons is very attached to the region. This has advantages as well as disadvantages. The main advantage is the presence in Lyons of very strong networks of people who have known one another for a long time. This greatly enhances and facilitates cooperation because they trust each other. The attractive scale of Lyons makes networking an effective weapon for business: much can be potentially realised by interpersonal relations. On the other hand, strong interpersonal networks are also a weakness, because the external orientation of the Lyon people is mostly too limited. This explains the shortfall in international relations of actors in Lyons. For high-tech and innovative activities in particular, where progress is made on a world scale, this is a threat. In the universities in particular, the international orientation falls short of the level required to participate optimally in international research projects and education programmes. This may be detrimental to the quality of the universities in research and education.

The *quality of life* is a fundamental asset of the Lyons region. Lyons rates high in different fields. Firstly, the location of Lyons near the Alps, not too far from the Mediterranean and only two hours from Paris, makes it an attractive place to live. Secondly, the city itself has much to offer in terms of cultural facilities, shopping areas, etc. Furthermore, the countryside is never far away. Also, compared to Paris, Lyons is not an expensive city in which to live. This explains why most firms in the region, in the health cluster as well, have little difficulty in attracting the (skilled) people they need.

The external *accessibility* of Lyons can be judged good. In particular the very frequent TGV-connection to Paris is important for the Lyons economy in general, but also for the health cluster, with its many contacts between Lyons and Paris. The airport offers access to ample destinations within Europe. For connections outside Europe, important for the large firms in the region, the situation is less positive. In most cases, intercontinental flights depart from Paris, but that is generally not regarded as a disadvantage. The external accessibility of Lyons will improve further in a few years, when the new TGV-track to Marseilles is ready. Travel time from Lyons to Marseille will then be reduced from three hours to one-and-a-quarter. This will also greatly enhance the attractiveness of Lyons as a place to live, as it brings the Mediterranean close.

6.2 Cluster-specific Conditions

As shown in section 4, the health cluster in Lyons consists of many different elements (hospitals, research, education and firms) and is of a large scale, if regarded at a high abstraction level. At a closer look, however, in some respects the cluster appears too fragmented. This holds in particular in the field of research. There are very many laboratories in different specialities, but they are mostly small-scale and therefore not cutting edge, or else they are very dependent on a few persons in the laboratory. A strong point is the presence of large firms in the region, such as the different Mérieux firms and others. They are leading firms in biotechnological fields and pharmacy and operate in international networks. However, in many cases they are governed from headquarters elsewhere in the world. Though these firms are rooted in Lyons by many research contacts and by tradition (Mérieux, Boiron), this poses a potential threat. Another conclusion is that the networks within the cluster are only weakly developed. In particular, linkages between the businesses in the region and the public research infrastructure are weak. Furthermore, the dynamics in the cluster are restricted because of the very low number of newly-created biotechnological or medical firms in the region.

6.3 Organising Capacity

Initiatives have been taken by various actors in the cluster to strengthen it in one way or another. One is the initiative taken by the universities to improve their contacts with industry, by setting up intermediary contract agencies such as ESUS and INSA-valor. Furthermore, one of the most important firms in the region, Mérial, together with Pasteur Mérieux Connaught, plans to set up a special course for higher education in virology and biology, to equip students with capabilities that are important for these firms.

A further case in point is the plan of the hospital organisation to concentrate hospitals in three poles and, more importantly, to stimulate cooperation in research fields that until now have worked separately from each other. The university helps the cluster by concentrating research efforts on a few core research themes. However, the policy of specialisation has been implemented only recently and its effects are not yet visible.

Though there are many public initiatives as well, in particular from the Communité Urbaine (see section 4), and a lot of effort and money is put in the cluster, some aspects of organising capacity could be improved. What is clearly missing in the organising capacity of the health cluster in Lyons is first, structural guidance for people who want to start a new firm in biotechnology or medical technology, and second, there is a lack of vision on the development direction of the focal poles in the city (Gerland, DOUA, Pole East, North and South). There is no spatial economic vision of the specific location requirements of each type of activity.

Although several initiatives have been taken to reorganise relationships and activities in the cluster in a better way to improve its functioning, much more needs to be done.

7 Conclusions and Perspectives

Thanks to growing welfare, an ageing population and technological developments, the contribution of the health sector to metropolitan economies will increase in the next few decades. The concentration of hospitals, pharmaceutical and biotechnological industry, medical business, and universities in Lyons form a sound basis for benefiting from this growth.

However, to benefit from longer term developments in the face of intensive international competition, a strong regional network is a prerequisite. Intensive regional interaction in the cluster should be considered a driver for growth and a stepping stone to a stronger international position. At this point, the situation in Lyons begs for improvement, as the level of interaction is rather low. From the cluster analysis, two main stumbling blocks to more vigorous interaction appear:

1 the separation between the public and private sector in research and development. Their functioning at cross-purposes, with some exceptions, has everal causes. First, the universities' culture gives priority to scientific publications: to adapt research to market needs is 'not done'. Second, it has been said that the professionalism at universities – methodology, quality procedures, international experiences – falls short of private sector requirements. The fact that the universities in Lyons participate in EU projects hardly at all is a signal of the isolated position. Third, the private sector has translated the growing international competition into a global strategic orientation; many companies are not aware of the fact that their home region can be a crucial asset to their own international position. The separation between public and private actors results in suboptimum functioning: opportunities for synergy are lost and facilities are sub-optimally used;

2 the low birthrate of companies in the field of health care. In the Lyons region, the number of start-ups in biotechnology, pharmacy, medical equipment and health care is very small. The low birthrate is due to several factors, as outlined in section 5.3. As a result, there are hardly any spin-offs and so no source for renewal in the cluster.

Breaking through the blockades requires a new orientation on the part of both universities (and public research institutes) and private companies. They have to be made aware of the mutual benefits they can deliver to each other and the impact of these benefits on their specific positions and the functioning of the cluster as a whole. For instance, the universities and public research institutes have to become aware of the fact that practical research raises the quality of their scientific output and the larger companies have to understand the advantage of start-ups for their innovation capacity and flexibility.

Table 7.13 gives an idea of the potential value that the various actors in the cluster can offer each other. On certain axes, such as universities-hospitals and hospitals-government, interaction is proceeding fairly well. However, on other axes, potential added value is not realised owing to a lack of contact and understanding.

A 'twinning' model could break down the walls between the actors in the Lyons health cluster and enable them to grasp mutual benefits for stimulating the cluster's functioning. Twinning is an organisational concept, developed in the computer industry in the USA and recently applied in the Netherlands by a Philips executive and the Dutch Ministry of Economic Affairs. It is aimed at facilitating start-ups through the larger companies. Large companies function as an umbrella for starting firms, to generate mutual benefits. The large company increasingly falls short of creatiing conditions for innovation and flexible market response, as they are organised to produce and market their products in the global arena. New enterprises can offer fertile grounds for creativity. In addition, they can facilitate the outsourcing of non-core activities by the larger companies. The larger ones can stimulate the development of start-ups with their knowledge of marketing, regulations and finance. Moreover, they can easily generate venture capital. The twinning model recognises the importance of a dynamic regional environment as a basis for international competition and as a complement to the growing 'footlooseness' of international business.

The twinning concept is especially attractive for Lyons, as the region accommodates several large firms in the health cluster, such as Mérieux, Rhône Poulenc and Boiron. However, the conditions for start-ups in the cluster are

Table 7.13 What the actors in the cluster can offer each other

Generates benefits for ...	Universities	Large firms	Start-ups	Hospitals	Government
Universities	X	Training and research facilities and human resources	Training facilities and advice	Human resources	Image, scientific output
Large firms	Funds for research and international networks	X	Advice, international networks and finance	Know-how on medicine	Image, tax income
Start-ups	Carrier perspectives for students, commercialisation of inventions	Flexibility and innovation power, testing ground	X		Image
Hospitals	Test and training facilities	Test facilities	Test facilities	X	Quality care
Government	Funds for research and education	Business sites and efficient regulations	Business sites, efficient regulation and finance	Business sites	X

yet rather unattractive and the need for research and development to enter into new markets is abundantly clear. Nowadays, the larger companies do not have structured programmes for facilitating new firms and spin-offs. Starters are too dependent on the informal goodwill of individuals in their environment. Besides ad hoc help of companies, the Chamber of Commerce assists new companies. However, the incubator facilities of the Chamber of Commerce are of a general nature and not equipped to assist business in biotechnology and health care. The Chamber has no special skills to facilitate activities in the life sciences.

In the case of Lyons, the twinning model should be extended to universities and public research institutes in order to integrate private and public research and private business development. Larger companies should not only support young firms, but also research programmes in their fields of strategic interest, for instance by outsourcing research to universities, supporting PhD students and opening their international networks to scientists. In their turn, universities should make themselves more attractive as partners for joint research with the private sector and as a stepping stone for business development. This implies, among others, upgrading the quality of research and collecting more knowledge on research procedures and licences.

The 'triad' of large companies, start-ups and universities also offers a opportunity to streamline research efforts and create scale economies in promising research directions. Further concentration and critical mass creation are needed for fundamental world class research. Even so, CNRS and INSERM should continue to play an important role in financing necessary fundamental research which is not directly commercially useful. The involvement of national research programmes can also counterbalance the role of private research, as this kind of research often is steered from abroad.

Developing the twinning model as a means to accelerate the interaction in the health cluster seems promising for the case of Lyons. However, it asks much from both the private and the public sector. The key ingredients seem to be leadership and vision. It is recommended that someone responsible for activating the health cluster along the lines described above should be appointed. This person should act as a broker between private firms and public research institutes, assist firms to articulate their research demands and scientists to formulate and improve their skills, and should work on a coherent vision and action programme to build the health cluster.

Note

1 Co-author: Dr Arjen van Klink.

References

Aderly (1995), *La Techerche à Lyon.*
Aderly (1996), *Lyon Technopole.*
Aderly (1998), *Lyon, Métropole de la Santé.*
Arthur Andersen (1996), *Étude sur le potentiel d'attractin de la région Lyonnaise en matière pharmaceutique.*
Assedic (1997), Salariés privés.
Berg, L. van den, J. van der Meer and P.M.J. Pol (1996), *Impact of the European Union on Metropolitan Cities*, Euricur, Easmus University Rotterdam.
Communauté Urbaine de Lyon (1998a), *Pour une Métropole technopolitaine 1998–2001*, Plan d'actions.
Communauté Urbaine de Lyon (1998b), *Mission Ville et Hôpital, Présentation, Actions réalisées et en cours.*
ERAI Entreprise Rhone-Alpes International (1998), *Réussir en Rhone-Alpes: les industries du genie biologique et medical.*
EZUS (1997), EZUS-Lyon 1: *Activité globale, fonctionnement, synthese de 6 ans d'activité.*
Hospices Civils de Lyon (1994), *Projet d'Établissement, Synthèse et Propositions.*
Hospices Civils de Lyon (1996), *Rapport d'Activité.*
Hospices Civils de Lyon (1997), *Hospices Civils de Lyon.*
Tableau de l'Économie Française (1997), 1988–99.
TEF (1997), Atlas du Grand Lyon, *La France et ses régions.*
Université Claude Bernard (1998), *Université Claude Bernard – Lyon 1 – secteur Santé.*
World Health Organisation (1996), *European Health Care reforms, analyses of current strategies*, WHO, Regional Office for Europe, Copenhagen.

Discussion Partners

Mr J.H. de Beauregard, Laboratoire Aguettant.
Mrs G. Berthillier, Institute des Sciences Cognitives.
Mr C. Boiron, Laboratoirs Boiron, Director.
Mr B. Brau, Merial, Vice President Manufacturing and Supply.
Mr C. Collombel, Dean of the Faculty of Pharmacy.
Mr B. Crouzet, Schering-Plough.
Mr Dechavanne, University Claude Bernard, Dean.
Mr A. Dittmar, research engineer, INSA and CNRS.
Mrs R. Eloy, Biomatech, Scientific Director.
Mr B. France, ARTEB (Agence Rhone-Alpes pour les Technologies Biomédicales).
Mr Fritch, L'Infirmerie Protestante.
Mrs F. Fuchs, Agence du Médicament, Manager.

Mr R. Garonne, Institute of Chemistry and Biology of Proteins.
Mr P. Girard, Dermscan, President.
Mr Y. Guyon, Chamber of Commerce, Department of Industry.
Mr C. Mantion, Rhône Mérieux, Distribution Pharmaceutics.
Mr H. Maupas, ARTEB (Agence Rhone-Alpes pour les Technologies Biomédicales).
Mr R. Maury, Aderly.
Mr R. Mornex, Hospices Civils de Lyon, Director.
Mrs L. Normand, Biomatech, Communcation Manager.
Mr E. Poincelet, Biovision, General Manager.
Mr A. Satgé, Chamber of Commerce et d'Industrie, Department of Economic Development.
Mr P.Y. Tesse, Chambre de Commerce et d'Industrie, Director.
Mrs C. Wantz, Communauté Urbaine de Lyon, Department of Urban Development.

Chapter Eight

The Cultural Cluster in Manchester

1 Introduction

The focus in this contribution is on the cultural industries in the city of Manchester. A recent paper from the European Commission estimated the European job-growth potential for the cultural sector at between 140,000 and 400,000 jobs a year. That is two-fifths of the European Commission's employment-growth objective. Manchester was one of the leading cities in the massive industrialisation in the eighteenth and nineteenth centuries but Manchester was also one of the first cities to experience the structural problems of de-industrialisation. The city acknowledged the cultural industries as an area for economic and job growth for the late 1990s and into the new millennium.

Although most observers agree that the cultural or creative industries contribute to economic growth, its very difficult to mark out the boundaries of these industries. Interpretations vary considerably but most observers include a broad range of activities. We have therefore followed the definition provided by the Manchester Institute for Popular Culture in their research of the sector in Manchester. We review the sector from a cluster perspective with some illustrations from the field of design, because there are many subsectors in the field of the cultural industries.

This contribution is structured as follows. First, section 2 sketches a concise profile of Manchester and its economy, followed by a discussion of the cultural industries in general in section 3. In the fourth section we put together the elements of the cluster in Manchester, including the firms, educational institutions, demand and the various government organisations. Next, section 5 describes the relations and dynamics in the cultural-industries cluster in Manchester. Section 6 puts the information of sections 4 and 5 in perspective and analyses the cluster according to our frame of reference that is developed in the second chapter of this volume. Finally, in section 7, we highlight some of the main conclusions and offer some suggestions for policy making.

2 Profile of the Manchester Region

2.1 Profile

The city of Manchester is the capital of England's North West region, the largest economic region in the United Kingdom (UK) outside London. The North West region's economy accounts for one-tenth of the UK output and is greater than the economy of, for instance, Finland or Denmark. The city of Manchester is by European standards a medium-sized city with just over 440,000 residents. The city is also the heart of the Greater Manchester area with a population of 2,578,000. The conurbation includes the cities of Manchester and Salford as well as the metropolitan boroughs of Bolton, Bury, Oldham, Rochdale, Stockport, Tameside, Trafford and Wigan.

Figure 8.1 Manchester in the United Kingdom

The city of Manchester is internationally known as the first city of the industrial revolution. The city has long been an important economic centre. From the time when a strong community of Flemish weavers settled in the city in the fourteenth century, it developed as a centre for textiles and related activities. The international scope of the city widened as soon as Manchester merchants got involved in European city partnerships such as the Hanseatic league. The city's history as a trading and production centre created the basis for Manchester's position at the very heart of the Industrial Revolution. Manchester was one of the leading cities in the process of industrialisation in the course of the nineteenth and twentieth centuries. As a result, Manchester's economic and political powers in the UK and also abroad was greatly enhanced.

Manchester had to cope with economic restructuring as the manufacturing sector started to decline from the late 1960s onwards. The city experienced severe problems, in common with many other major industrial cities on the European mainland and in the US. Between 1975 and 1990, Manchester lost over 100,000 manufacturing jobs and approximately one-third of its population. In the same period the service sector created many new jobs (90,000) but the problem was that most of these new jobs did not go to the people displaced by the job losses in manufacturing. As a consequence, Manchester's growth areas in the city centre and around the airport to the south of the city, are both at walking distance, of two of Manchester's worst areas for poverty and unemployment (Carter, 1997).

2.2 The Momentum for New Policies

For various reasons, the policy context for economic development has changed drastically in Manchester in the 1990s. For decades Manchester's city government had been dominated by the Labour Party, which held between 80 and 85 per cent of the seats in the city council. This ensured a stable political climate. However, change occurred also within the local Labour Party. An important moment was the defeat of the Labour Party in the national elections in 1987. Many people in Manchester's political circles had expected Labour to win and counted on direct financial support from the national government for intervention to overcome the difficult economic situation of the city. That hope disappeared with the third term for a Tory Prime Minister in Downing Street. The political situation compelled the city government to think of a different strategy to overcome the serious economic problems. Up to the early 1990s, cities in the UK had no statutory obligation to publish formal economic development strategies. This changed in 1989 when the national government brought in such a requirement and in 1992 Manchester published its first Economic Development Strategy.

Manchester's application for support from the European Objective 2 programme also required the city to develop such a strategy, while the City Challenge policy introduced in 1992 required local partners to come up with a wider vision for their area as well, linking programmes, projects and resources in a strategic way (Parkinson, 1998). All these factors together created the momentum for a new strategy. The vision and outline of the strategy were laid down in the first City Pride document in which Salford and Trafford joined with Manchester and to which the private sector contributed as well (City Pride Partnership, 1994). City Pride is not considered to be a 'traditional' policy document for the city but is treated as a manifesto for the city. The manifesto promotes the development of Manchester as:

- a European Regional Capital, a centre for investment growth;
- an international city of outstanding commercial, cultural and creative potential;
- an area distinguished by quality of life and sense of well-being enjoyed by its residents (City Pride Partnership, 1994; 1997).

Recently the City Pride document was revised to reflect a number of changes: Tameside's decision to join the City Pride partnership, a new (Labour) government in Downing street, and the advent of the National Lottery funds. The most dramatic event to affect the city's strategy was the IRA bombing, which severely damaged part of the inner city's shopping area in 1996. Although this seriously damaged the economy of the city centre in the short term, it did create an opportunity for the renewal of the area, including investment in cultural facilities.

2.3 The Greater Manchester Economy

The greater Manchester economy has also undergone significant changes in the past decades. In the 1960s, the manufacturing sector prevailed in the Manchester economy, employing half of the working population. Nowadays, the manufacturing sector accounts for just over 21 per cent of employment (see Table 8.1).

Manchester is an important regional education centre with four universities (see section 4), an asset that has developed into a significant economic feature. Another great asset for the area's competitiveness is Manchester Airport, which offers excellent national and international connections to 175 destinations with scheduled flights. It also induced employment growth in the last few years and continued employment growth is forecast in the years to come (9 per cent to 2002; Manchester TEC, 1998). The airport has also attracted new economic investment nearby.

Table 8.1 Employment in the Manchester TEC area for 1995

Sector	Employment	%
Agriculture and fishing	230	0.0
Energy and water	4,650	0.9
Manufacturing	81,030	16.0
Construction	18,860	3.7
Distribution, hotels and restaurants	102,170	20.2
Transport and communications	41,530	8.2
Banking, finance and insurance, etc.	104,140	20.5
Public administration, education and health	136,480	26.9
Other services	17,860	3.5
Total	506,950	100.0

Source: AES (ONS) 1995, in Manchester TEC, 1998.

In general, the economic performance of the Manchester TEC area (measured in Gross Domestic Product (GDP)) remains below the average for the North West region as well as the UK average (see Table 8.2). Investment levels in the TEC area are well above the national average, mainly because of the major (re)construction projects in the city triggered by the IRA bombing.

Table 8.2 Economic performance of the Manchester TEC area in perspective

	Annual average change in GDP		Annual average change in investment	
	1988–98	*1998–2000*	*1988–98*	*1998–2000*
Manchester TEC	1.2	2.5	3.2	3.5
Northwest region	1.4	2.4	2.6	3.3
United Kingdom	1.8	2.9	0.5	3.2

Source: CE/ER LEFM, 1997, in Manchester TEC, 1998.

Unemployment in the Manchester TEC area[1] is estimated at 6.5 per cent, well above the UK average of 5.0. However, unemployment in the TEC area has dropped by one-quarter over the past year, following the national trend of lower unemployment rates. Unemployment across the Manchester TEC area varies among the districts, with the city of Manchester accounting for the highest rate of 10.6 per cent. By a new measure of unemployment about to be adopted by the British government, the rates would be considerably higher.[2]

The unemployment rate of Manchester TEC area would be 9.8 per cent according to the new approach, with the city well above that rate.

The economic outlook for the Manchester area is inextricably linked to national and international economic development trends. On the one hand Manchester holds several trump cards with its airport, educational infrastructure and many of successful major employers (Manchester TEC, 1998). On the other hand, the levels of unemployment and poverty are high and might jeopardise Manchester's sustainable economic development. Many in the Manchester area view the city's successful bid for the Commonwealth Games as an opportunity to trigger additional investment in the area, present the area to an international audience and another vehicle to strengthen the public-private partnerships.

2.4 The Cultural Industries

Manchester has seen important changes in the past. Manchester was not only at the forefront of the industrial revolution, it was also a centre of innovation, research and invention. It was at Manchester University that the first computer with a stored programme was developed (1948). The city's social, economic and cultural base has also provided the stimulus for cultural innovation. Manchester is also known to be at the cutting edge of popular youth culture, more particularly in the field of music and design. In the late 1970s/early 1980s Manchester pop acts[3] represented a vibrant pop culture in the city, which reached an international audience. In the second half of the 1980s, Manchester transformed itself into the centre of 'dance culture', with world-famous dance clubs such as the Hacienda bringing 'underground culture' to a larger international audience. At the time of writing, Manchester-based Oasis is one of the most successful bands in the UK.

In the last few years the city has recognised that there is economic potential in an area that is not directly visible in the traditional economic figures, in its cultural industries. Manchester TEC has included the *arts, culture* and *media* amongst the 14 key sectors in the TEC area and credits 2 per cent of the TEC area's employment to that sector (Manchester TEC, 1998). According to their figures, the sector was quite stable in the early 1990s and holds promise for the future. More recent extensive quantitative research in the Manchester Training Enterprise Council Area has adopted a wider view on the cultural industries and estimated that 3.56 per cent of the working population (18,058 jobs) is (self)employed in cultural enterprise (MIPC & DCA, 1998). In comparison with other UK cities, Manchester's employment figures stand out positively. Only smaller towns with a few large employers (Cardiff) do better (see Table 8.3). Some questions suggest themselves. What is the significance of the cultural industries? What are the main characteristics of the cultural industries cluster in Manchester?

Table 8.3 Manchester cultural enterprise in UK perspective

	Share of employment
Manchester	3.56%
London	2.3%
UK average	2.3%
Cardiff	4.3%

Source: MIPC & DCA, 1998.

3 The Cultural Industries and Design

Interest in the economic impact of culture has increased in the 1990s. For a city, culture is an inseparable part of the set of amenities and services offered to their community, making it an important location factor for people and business. Increasingly, culture is part of the urban revitalisation strategies for degraded urban areas. The supply of cultural facilities can also stimulate the creative base for other sectors in the city. Figure 8.2 illustrates the economic impact of culture.

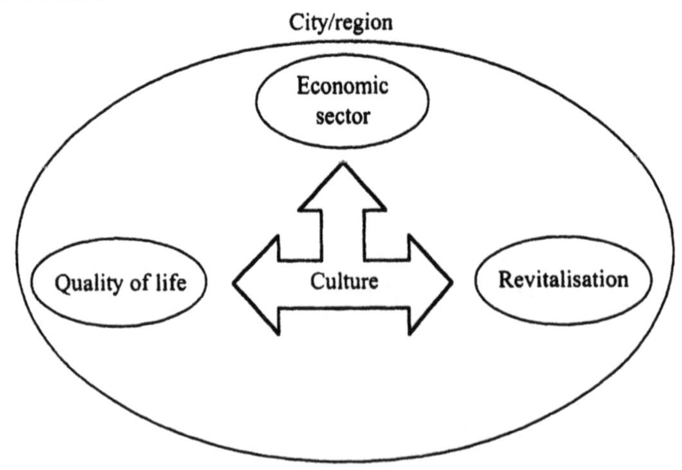

Figure 8.2 The economic impact of culture in a city or region

The European Commission has identified the various sectors of the cultural industry as a major economic and social force in Europe. Employment in the culture and crafts sector is estimated to account for 2 per cent of overall employment in the European Union. The growth of cultural employment has been strong in the past 10 years, exceeding general employment growth figures. Table 8.4 presents some examples. A number of European cities rate the growth potential of cultural employment high (EDURC, 1997).

Table 8.4 Examples of cultural employment growth in some countries

Country	Employment growth rate	Growth period
Spain	24.0%	1987–94
France	36.9%	1982–90
United Kingdom	34.0%	1981–91
Germany	23.0%	1980–91

Source: European Commission, 1998.

Most observers would agree that there is growth potential in culture and the cultural industry. The demand for cultural 'goods and services' has been rising because of interlocking social and economic trends:

- the growing importance of the service sector has led to changes in life style, higher income and an increase in the proportion of that income going to culture, leisure and entertainment (3 per cent on average);
- people live longer and the over-55s in particular consume more cultural goods and services with their increased leisure time and disposable income;
- rising standards of education have a positive effect on the demand for culture;
- on average the free time of European citizens has increased;
- the participation in cultural life has become more diversified;
- growing urbanisation reinforces the observed correlation between the supply of culture and the degree of urbanisation (European Commission, 1998).

In other words, the market for cultural consumption has expanded and the customers have more time available for it and are willing to spend on it. Opinions on the demarcation of the cultural industries are more divided. The interpretation of what is to be included differs among the various European countries. In the European Commission's efforts to assess the employment impact of the cultural industries, a distinction was made between collective and individual cultural behaviour. Heritage, music concerts, the performing arts (including festivals) and cinema are seen as collective cultural activities. Individual cultural activities are mainly for consumption at home, such as television, radio, reading and libraries (including literature) and recorded music. In addition there are many new forms of cultural production and consumption due to the seemingly unlimited possibilities that new information and communication technology (CD-ROM, Internet) makes available. Another classification (MIPC & DCA, 1998) incorporates craft, design, performance, visual arts, music, photography, heritage, multimedia, film/media, advertising, architecture, authorship and education.

It is very difficult to produce an undisputed definition of the cultural industries. Where does it fit in the traditional economic classifications? Where do we draw the line between cultural activities and leisure? On the fringes of the cultural sector there is a whole series of economic activities (for instance, electronic equipment such as television sets or hi-fi) in which employment is somehow linked to the wealth and distribution of culture (European Commission, 1998). Culture is also a major factor in urban tourism (Russo, 1998).

We feel that these cultural activities have become increasingly important for the urban economy but that they sit uneasily with the sector-based economic classifications, comparable with many other urban economic growth clusters.

The key word for culture and the various sectors of the cultural industry is creativity. No wonder that sometimes the notion of the creative industries is put forward. Creativity is the engine for the production of cultural goods and services (works of art, movies, music, clothing, etc.). These goods and services are distributed (auction, record store, museum) to the consumers. The supply chain to which the cultural activities are central is depicted in Figure 8.3. In the context of the digital revolution, that sequence is not as straightforward as it used to be.

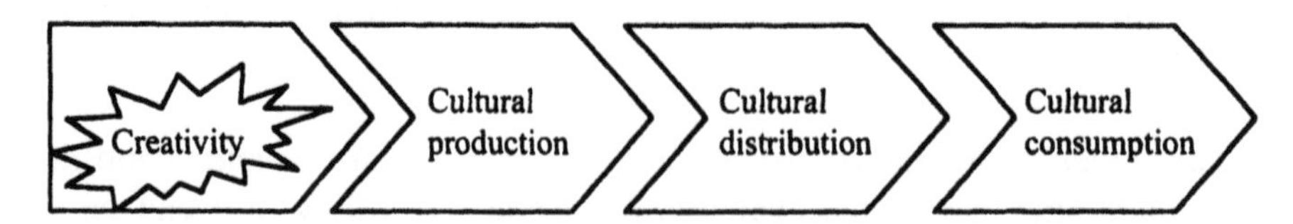

Figure 8.3 Example of a supply chain in the cultural industries

Another important feature is the spatial scope of the cultural industry supply chain. The present society, with its technological possibilities and increasingly global markets, stimulates the internationalisation of the distribution and consumption of culture. The music industry is a good example of an industry where the main distributors (namely record companies) have become international conglomerates with decision-making powers concentrated in the global cities. Popular music (predominantly American and British) is consumed all over the world. One could say that cultural production and creativity are also affected by internationalisation but have a stronger attachment to the local level.

In many cases activities that are considered parts of the cultural industries (for instance architects, musicians, designers) are parts of other supply chains as well. In the case of design, typically, a designer would team up with a manufacturer and the latter would take the designer's idea into production (see Figure 8.4). The digital revolution has impacted here as well. Nowadays, the supply chain is no longer formalised. Figure 8.4 illustrates that design activities can be part of various supply chains. Designers are much more entrepreneurial and self-supporting. For example, in the music industry, artists and composers can make hit records from their bedrooms.

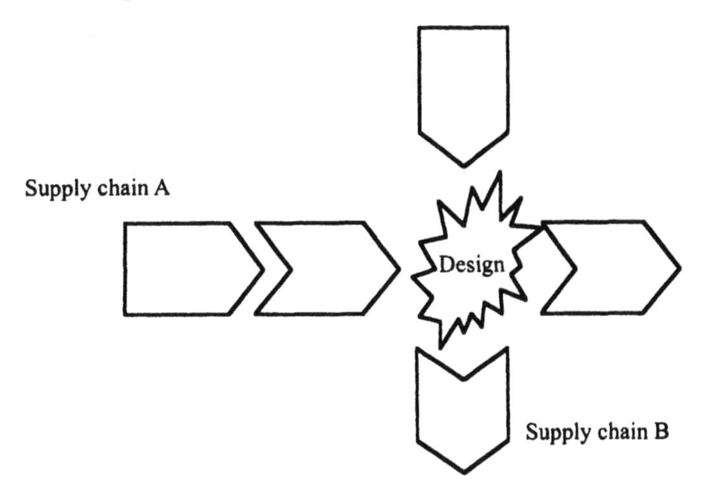

Figure 8.4 The example of design

In sum, the example of design illustrates that many subsectors of the cultural industries are intrinsically connected with many other sectors in the urban economy. That makes these industries difficult to grasp, but at the same time highly relevant to the urban economy.

4 The Cultural Industries Cluster in Manchester

4.1 Introduction

In this section the principal actors in the field of the cultural industries are described. We will try to present a general picture of cultural enterprise but our particular interest involves the subsector of design. The previous section illustrated that the cultural industries can make a significant contribution to city employment, although it is difficult to define the boundaries of these 'industries'. We have to take the definition of the Manchester Institute for Popular Culture (MIPC) in their research commissioned by the city of Manchester. First, we will discuss private enterprise. Second, we will turn our attention to the demand for the goods and services of cultural enterprise. Third, we consider the educational institutes that are relevant to the cluster. Fourth, we will address the policies of the city administration and other (semi-)public actors that are of relevance for the cluster. Finally, we will discuss two examples of spatial clustering in specific parts of the city.

4.2 Private Enterprise

Cultural sector enterprises in the Manchester TEC area are categorised into 12 subsectors: craft/design, performance, music, visual art, photography, heritage, multimedia, film/media, advertising, architecture, authorship and education (see Table 8.5). Together these activities account for 3.56 per cent of employment (MIPC & DCA, 1998). The researchers claim that this is the minimum size of the sector in the city. Contrary to popular belief, only 5 per cent of investment in cultural enterprise comes from public sources.

At first sight, the size distribution of cultural sector enterprise in Manchester is surprising. The figures contradict the general idea that cultural businesses are relatively small with sole practitioners and small and medium sized enterprises (SMEs) setting the scene. In the case of Manchester the reality is different; the share of sole practitioners and SMEs is far below the UK average and the large enterprises (LEs) account for more than half of employment. One factor influencing this deviation from the national and European average is that the MIPC's definition of the cultural industries covers subsectors which are not conventionally included, such as press, advertising and architecture. If these subsectors are excluded a more familiar pattern arises (see Figure 8.5). Even so, large companies are more plentiful in Manchester than in other cities.

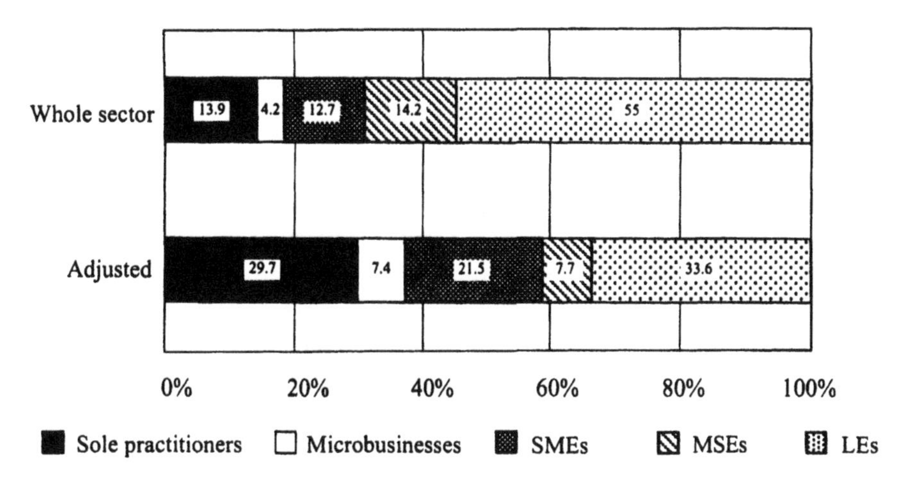

Figure 8.5 Size distribution for cultural enterprise

Source: MIPC & DCA, 1998.

The research confirms that Manchester has strength in the subsectors of design, music, architecture and media. Manchester is relatively strong in those areas for historical reasons. Table 8.5 shows the distribution of employment across subsectors in the cultural sector in Manchester.

Table 8.5 Employment and firm distribution in the subsectors of the cultural sector

Subsector of the cultural sector	Employment %	Share of cultural enterprise %
Craft/design	12.8	25.7
Performance	5.1	8.5
Music	8.8	25.7
Visual Art	3.5	8.3
Photography	2.0	3.3
Heritage	5.8	0.8
Multimedia	4.7	1.8
Film/media	19.7	6.9
Advertising	2.8	5.1
Architecture	12.8	4.7
Authorship	16.0	8.8
Education	6.0	~0
Miscellaneous	–	~0
Total	100%	~100%

Source: MIPC & DCA, 1998.

Design One of the principal subsectors in Manchester is craft/design. The subsector accounts for 12.8 per cent of cultural-sector employment and just over a quarter of all cultural businesses. It is a growth sector for the UK in general and for Manchester in particular. Design is concerned with the development and production of 'symbolic knowledge'. Designers make products, clothing, graphic, interiors, etc. that appeal to people. The city has a reputation for its 'cutting edge' designer community with some (inter)national 'movers and shakers'. They work in such fields as fashion, graphic design and property development. The design industries are characterised by a large number of SMEs and sole practitioners. Some Manchester-based firms have gained wider recognition in the UK or even abroad. The field of design has almost by definition many linkages with other subsectors of the cultural industries.

Music Manchester is famous for its history in popular music. Since the first half of the 1960s Manchester has produced a sequence of successful popular bands. The city has always had a special local music scene that has renewed itself many times in the past. More than once, Manchester has reached the headlines in the world of pop and rock. Several times the sound and success of Manchester-based acts and bands turned into a 'Manchester phenomenon'. The popular music business is a truly global market in which the UK has a strong tradition. Manchester's music scene thrives on its creative base, not on the presence of any of the major decision-makers established in London, New York or Tokyo. It is, therefore, not surprising that the firms are relatively small in Manchester. The vast majority of music businesses in Manchester are sole practitioners. Nevertheless, the music business in Manchester accounts for 25.7 per cent of cultural enterprise and 8.8 per cent of cultural sector employment.

(Broadcasting) media The broadcasting media sector is deeply rooted in the greater Manchester area. The BBC keeps an office in the city, but more important is the Granada Media Group. Granada is the largest company in the independent television network, ITV. Granada is ITV's longest established broadcasting firm and its programmes date back to the first day of transmission in the mid-1950s. Granada is Britain's largest commercial programme maker, broadcasting to northwest England seven days a week and contributing programmes to the national independent network. The broadcast of Granada TV's programmes is destined for a large regional audience, but at the same time Granada productions have wider national and global distribution including the world's longest-running, highest-rating drama serial, *Coronation Street.*

Apart from Granada and BBC, a number of smaller firms are making programmes. The typical small firm employs five people in the permanent staff supplemented by flexible employees depending on the projects in hand.

Multimedia Cultural-sector enterprise in the field of multimedia accounts for only 4.7 per cent of employment and 1.8 per cent of the number of firms. Nevertheless, it is growing in the city and opens prospects for the future. At present most of the multimedia companies in Manchester could be characterised as being part of the IT or more traditional broadcasting sectors which means that the overall level of employment and the number of enterprises which can be classed as 'multimedia' may be underestimated. As the multimedia supply chain develops, however, an increasing number of individual entrepreneurs and micro-businesses are emerging as subcontractors to the larger content producers. This means that the total size and share of employment which can be attributed to multimedia is growing and will continue to do so.

Special 'cultural distributors' An important role in the cultural cluster is played by institutions that are primarily distributors of culture in general and cultural goods and services in particular, namely, galleries, museums, art houses etc. In Manchester there are numerous places and institutions that are important 'points of distribution'. Manchester City Art Gallery, for example, has more than 50,000 works of art in its collection. The Gallery is closed until 2001 because of a major expansion and (multi-million) upgrading project. The renovation is assisted by a grant from the Heritage Lottery Fund and support from Manchester City Council. Another cultural focal point in the city is Cornerhouse, a centre for cinema and visual arts. Cornerhouse is situated in the city centre and comprises three cinemas, three galleries and two bookshops, a popular bar, a café and a cappuccino bar. Cornerhouse is the largest regional arts centre in the UK, and it is the only place in Manchester where people can enjoy cinema and contemporary art under one roof. Every year the three cinemas at Cornerhouse present over 4,000 screenings of releases and the three floors of galleries with contemporary arts are free to the public. A new development is the opening of the CUBE Gallery on Portland Street as a centre for urban design.

There are also small-scale initiatives such as the Castlefield Gallery, an artist-run organisation offering artists studio space as well as a programme of exhibitions and education events. The organisation also acts as an agency for public and private clients seeking to commission artists (over 300 practitioners are in their database) to create permanent site-specific works of art.

4.3 The Demand Side

An important feature of the cluster is the demand for cultural goods and services. How is the demand for these goods and services structured? The cultural firms supply to many different customers. First, we need to distinguish between the local and regional markets on the one hand and national and international markets on the other hand. The local demand is very important for Manchester: the city is the cultural centre of the greater Manchester area, with over 2,500,000 people. Some of the Manchester-based firms sell their goods and services (inter)nationally. Another important component of the demand side is whether cultural goods and services are delivered to the end users or to other firms in the region along other supply chains (as indicated in section 3). Finally, there are differences between the many subsectors of the cultural industries as well. Some of the subsectors in the cultural industries operate in truly global markets, for instance popular music and design, against fierce global competition.

4.4 Education

Education institutions are significant actors in the field of the cultural industries. Manchester is home to a very large student population. More than 70,000 full-time and 100,000 part-time students are enrolled at one of the four universities in the greater Manchester area: the University of Manchester, Manchester Metropolitan University, Salford University and University of Manchester Institute of Science and Technology (UMIST). The Manchester-based universities are located close together in the largest university quarter in the UK.

Table 8.6 Students at Manchester institutions

Institutions	1994/1995	1997/1998
Manchester Metropolitan University	28,381	28,362
Victoria University of Manchester	21,000	23,571
Salford University	8,304	18,603
UMIST	7,063	6,780
Total	64,748	77,316

Source: MIDAS, 1999.

The first three universities in Table 8.6 are of particular interest for the cultural industries. Victoria University of Manchester is the oldest amongst the city's universities and offers students a traditional Arts degree. Both the Manchester Metropolitan University ('Met') and Salford University offer programmes that are more tailored to the various fields of cultural industries. Manchester Metropolitan University is one of the new universities created when the British government introduced one system of higher education in 1992. In comparison with the traditional redbrick universities, the 'Met' has a more vocational attitude. The 'Met' offers various programmes (architecture, 3D design, fashion/clothing) in the faculty of Art and Design. The Faculty at the Crewe and Alsager campuses in Cheshire also provides a programme of arts, design and performance (Manchester Metropolitan University, 1998). Salford University has a Faculty of Arts Design and Technology and a Faculty of Media, Music and Performance. The Royal Northern College of Music is one of the leading conservatoires in the UK. There are several colleges of further education (FE), with a significant number of arts-based courses.

4.5 Government Policies

Manchester has given higher priority to the cultural sector since the City Pride initiative in which commercial, cultural and creative potential were key elements. The city has provided opportunities for the cultural sector to contribute to urban regeneration. Culture has an integral role within the city's regeneration strategy as an significant source of employment, a contributor to quality of life and as a location factor. Manchester City Council published an arts and cultural strategy for Manchester to which the use of its cultural heritage, the promotion of cultural quarters, participation in cultural activities and the increase of the economic benefits are central.

The cultural industries cluster has developed without direct public sector support. The Manchester Training and Enterprise Council (MTEC) provides for business start-up, offering packages of training and business advice. These programmes are also open to cultural enterprises, but the experience of these entrepreneurs is that the programmes are not tailored to their needs. MTEC has recently developed some pilot programmes targeted at the media. Manchester Technology Management Centre (MTMC) is also active in the field of business support. The centre is part of the Manchester Institute for Telematics and Employment Research in the Faculty of Humanities and Social Science at the Manchester Metropolitan University. MTMC offers a range of services to the region's small businesses and community firms, including technology transfer. It promotes the use of information and communication technology and has knowledge in the field of electronic commerce (MTMC, 1999).

The City Council does not pursue many explicit policies that aim to support the cultural enterprises directly. Based on the research commissioned by the MIPC, a strategy has been developed to target support to the cultural industries sector. This has been developed within the context of the City Pride initiative in partnership with MTEC, the North West Arts Board and MIDAS. The strategy proposes a number of measures to promote the development of the cultural industries as an area for future job growth. The aim is to provide core infrastructure for support to the cultural industries sector across the City Pride area through:

- the provision of specialist advice and support to cultural industries;
- support for networking and collective initiatives for which key targets are design industries (fashion, crafts, graphic, product and interior design), music and media;
- provision of marketing support for cultural industries SMEs;
- encouragement and development of multimedia skills and products within the cultural industries sector;
- improving skills and knowledge to meet the needs of SMEs.

This new City Council strategy has to be delivered through partnerships with organisations in the area of business support such as Manchester TEC and the MTMC.

The National Lottery Fund Although the City Council had a new favourable strategy lined up, it had very limited resources to invest in arts and culture. Manchester was able to attract investment in the city's cultural infrastructure through the National Lottery Fund. The National Lottery was established by parliament to raise money for worthwhile causes. The net revenues (some 28 pence out of every pound spent) from the National Lottery are allocated to six good causes – arts, sports, charities, heritage, celebrating the millennium and health/education/environment. The Arts Council of England, Scotland, Wales and Northern Ireland Heritage Lottery Fund distributes the budget for arts and culture amongst the many applications from cities and all sorts of organisations. In 1997, £1.7 billion was handed over to the good causes and over £660 million paid over to the government in tax by Camelot, the company running the National Lottery. The renovation and expansion of the Manchester City Art Gallery is funded by that initiative and Cornerhouse has also applied for funding to expand its complex.

North West Arts Board The North West Arts Board (NWAB) is the regional arts development agency for Cheshire, Greater Manchester, Lancashire, Merseyside and the High Peak of Derbyshire. It is one of 10 regional arts boards in England funded by central government and part of the national arts funding system. The Arts Board supports the arts of the northwest region and provides a range of advice, advocacy, finance, information and other services by itself or in collaboration with partners in the region (public, private). The provision of core funding for the region's arts organisations is still a very important function, though the focus has shifted somewhat from a traditional funding agency for the arts, to creativity stimulated also by the use of new media. The Arts Board takes an active role in advising and encouraging applications to the National Lottery Arts Fund.

4.6 Important Locations within the Design Cluster

Cultural private enterprise is spread throughout the city and region; nevertheless, there are two areas that symbolise the cultural industries in the city, in particular for the design sector. First, the Castlefield area is home to some of the more 'mature cultural enterprises'. The Castlefield area is the site of a Roman outpost (Mancunium). In the 1700s and 1800s three major canals were joined there, and important rail connections crossed it. From the early 1900s it fell into decline but in the last few years it has been restored to become part of the cultural heart of the city and preserving the typical nienteenth-century industrial architecture. It now includes a mix of residential, cultural, commercial, leisure, environment and community activities.

Second, the Northern Quarter symbolises the cutting edge of many subsectors of the cultural industries. The Northern Quarter is home to over 550 businesses and organisations. It is a place for shopping, music, food and drink, entertainment, fashion, living and working. For instance, there are four design outlets for alternative fashion within walking distance from one another, which offer high-quality clothing from local designers. The area has numerous fashion outlets, cafés, pubs, bars, restaurants and clubs, 20 music stores and specialised arts, crafts and jewellery shops. The regeneration of the Northern Quarter is supported by a group of enthusiastic local entrepreneurs who have formed the Northern Quarter Association representing the local commercial and residential community (see Figure 8.6).

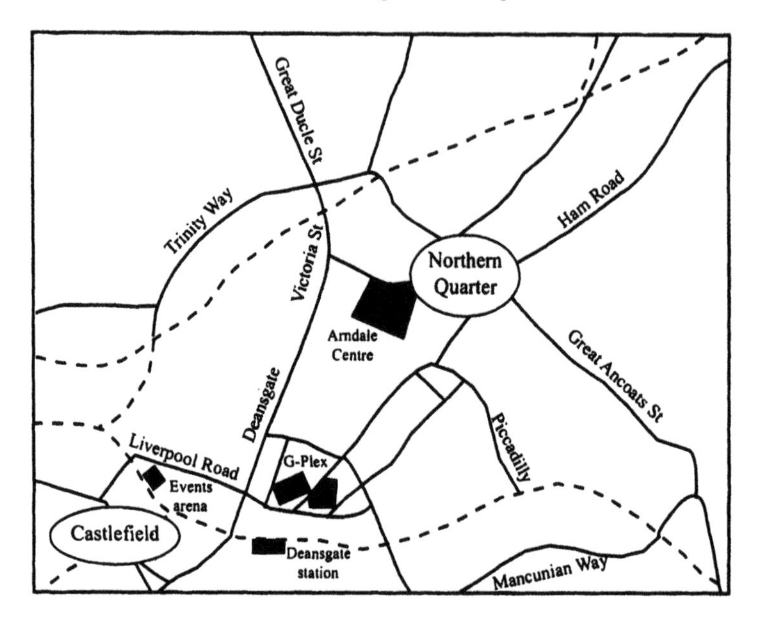

Figure 8.6 Locations within the design cluster

5 Dynamics in the Manchester Cluster

This section is dedicated to the interaction and dynamics in the cultural cluster in Manchester. In an environment where creativity is vital, strategic interaction between different actors in the cluster is of great importance for the competitiveness of the cultural cluster. It is in this light that this section describes and analyses some of the crucial relationships in the cluster as described in section 4. Figure 8.7 depicts some of the categories of relationships in the Manchester cluster. The paragraph numbers correspond to the numbers in the figure.

5.1 Relations amongst Cultural Enterprise

From section 4.1 the average size of the cultural businesses is smaller than that in many other sectors. The sector is not characterised by high degree of organisation. It is both informal and flexible. Although there is little formal structure, the sector is highly networked and interlinked through many informal contacts at the same time. Cooperation is essential in a sector with many small firms. However, the awareness of the city's supply of cultural products and services is low. That firms in the field of cultural enterprise need to network is clear from the example of the design companies. Clearly, these entrepreneurs prefer low cost accommodation, while the 'cultural feel' of its environment is also important. There is a strong tendency to cluster in specific parts of the city, e.g. the Castlefield, Northern Quarter. The proximity of firms of the 'same kind' is another important location factor. Nevertheless, the cluster does not comprise many large enterprises that could pull the cluster together and offer international contacts for the SMEs.

5.2 *Interaction between the Cultural Enterprises and Education: the Example of Design*

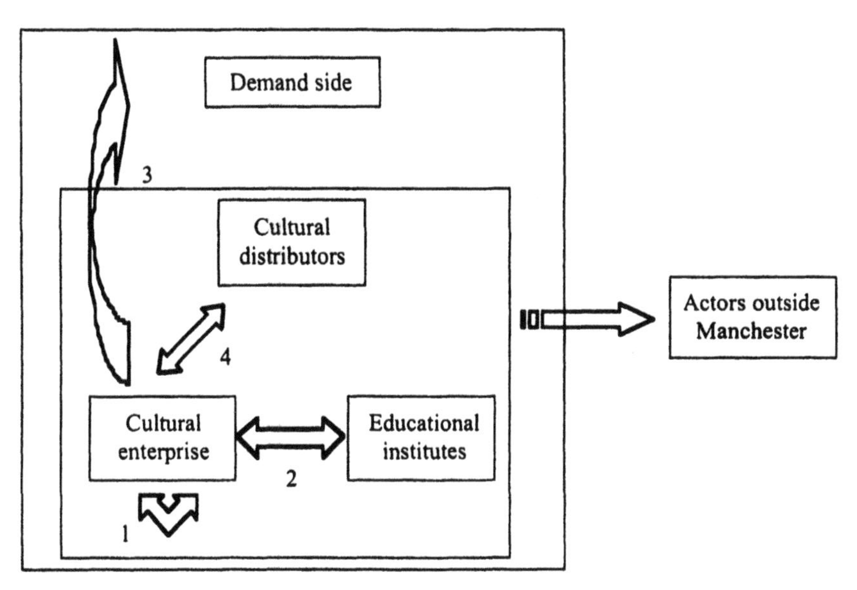

Figure 8.7 Relations and dynamics in the cluster

The linkages between the design and cultural business in general on the one hand, and the education institutes on the other are generally underdeveloped, although there are differences amongst subsectors. In the field of media, Granada has good relations with Salford University. Generally speaking, universities have difficulty to keep up with the rapid changes in the subsectors of the cultural industries. This is not just a problem exclusive to Manchester. The feeling is that in respect of culture and especially design, modernisation of higher and further education is needed. Training programmes are out of touch with the practice of cultural enterprise and work-placement schemes are in many cases not part of the curriculum.

There are differences also between the universities. The former polytechnics, which have been part of the new university system in the UK since 1992 (Manchester Metropolitan University), have a more vocational attitude than the classical Victoria University of Manchester.

Potentially, the universities could play a role as business incubators. Indeed the student population of Manchester is an important economic asset. Manchester has one of the largest university campuses in Europe. From these incubators, designers (for instance) could sell their cultural goods and services to the local demand in the city. None of the universities have as yet outlined a strategic policy aiming at such a role for the cultural-industries cluster.

5.3 Interaction between Cultural Enterprise and the Demand Side: the Example of Design Industries

Demand for cultural goods and services comes from a number of directions. Section 3 has already illustrated that many subsectors do not fit within very formal and standardised supply chains. Therefore a firm could be your partner today, your competitor tomorrow and your customer the day after tomorrow. Section 4 pointed out that there are many different aspects to the demand side as well. An interesting example of the *business-to-business* market is Granada TV. Granada TV produces 30 per cent of ITV's UK-commissioned programme hours, while it is only 12 per cent of the ITV system. Granada sells its products all over the UK, and in case of very successful programmes, all over the world. Granada is committed to contracting out one-quarter of its output to independent firms, making it an important customer for the SMEs in the Greater Manchester area. Many of these SMEs have been set up by former Granada employees. It is also clear that the outlook for further growth in the traditional media sector is constrained by the attraction that London exerts on media businesses throughout the UK.

The majority of the designer industries serve the *regional market*. For some of these firms the *UK* (read London) and *overseas market* is becoming more important. In same cases this is caused by specialisation in a small market niche; in other cases there is a perception that British industry in general is not too keen on incorporating (British) design in their production process. The subsector of fashion is a good illustration for Manchester. The city has a large fashion designer community on the one hand and there are still some clothing manufacturers (although far less then in the previous century) on the other. These sectors operate in different business environments that have no connections to each other.

5.4 Interaction between Cultural Enterprise and the Special Cultural Distributors

Often firms or organisations operating as 'distributors of culture' are part of what is considered to be the cultural industries. We do not want to exclude them from cultural enterprise; it is important to emphasise the role of organisations such as Cornerhouse, but also other galleries, shops, museums etc. One of the main problems of Manchester's cultural industries is their lack of visibility for the business community in general but also for policy makers. For example, Manchester is well known for its creative firms but only recently the economic potential of the cluster has been recognised. As a result a large-scale investigation into employment in the cultural industries has been commissioned.

The distributors of culture symbolise the cultural scene in Manchester and could play a stronger role in making the cluster more visible than it is today, all the more, since the cluster does not have too many larger firms (such as the big pharmaceutical firms in Vienna's health cluster or Nokia in the telecom-cluster in Helsinki) that could be flagships for the cultural enterprise. Such flagships could help the cultural enterprise to become better known and capture more of the regional demand, of which a portion is still leaking away to London.

5.5 Dynamics: Made in Manchester

An interesting development was the introduction of the new logo to market Manchester. The slogan and logo were dismissed by a group of cultural entrepreneurs who wanted to create an association with quality, creativity and innovation. The city is home to a very creative and active cultural enterprise cluster, that could have played an important role in a much earlier stage of the campaign. Moreover, involving the local cultural entrepreneurs would have also helped to increase the visibility of the cluster.

6 Confrontation with the Frame of Reference

Having extensively described the actors and relations in the cluster, this section puts it in the general research framework as outlined in Chapter Two. The framework summarises the spatial-economic context in which the cluster functions, the different aspects of organising capacity and the cluster development.

6.1 Spatial and Economic Context of the Cluster

The cultural industries, and design in particular fulfil a function within the Manchester economy, which has undergone a major transformation. The city suffered a spiral of decline as the preponderant position of the manufacturing sector faded away. Nowadays, the city's transformed economic structure is very different from that of the early 1970s: service-based activities provide most of the jobs in the greater Manchester area.

Much has improved in Manchester for the good in the last 10 years. The supply of cultural and other leisure facilities has increased. The attitude towards renewal and innovation – 'the cultware' – has changed for the better. The spirit is reflected in the vision laid down in the City Pride documents, in the successful urban regeneration projects, in the city's closer involvement with the European Union and in the city's raised cultural profile.

6.2 Cluster Development

Indeed, culture has played an important role in the regeneration process in Manchester. The role of the cultural sector is diverse. Culture and the cultural institutions are part of the mix of amenities that make the city more attractive to citizens, the business community and visitors. Cultural innovation within the city has strengthened the city's creative base, with a potentially positive effect on other economic sectors. Last but not least, the developments in the field of private cultural enterprise signal that it is also an economic sector that contributes to economic and employment growth in Manchester. Research has revealed that these cultural or creative businesses are an economic factor to be reckoned with and hold promises for the future (MIPC, 1998). These cultural firms draw on Manchester's history and reputation in media and popular music and have developed without direct public sector support.

The 'cultural industries' cross the traditional boundaries of economic sectors and there are many activities that could be included within such as craft, multimedia, film, architecture and advertising. Design is a key subsector within Manchester's cultural business, it accounts for one quarter of the cultural business. Design of buildings, interiors, clothing, ceramics, films, magazines, CDs, etc., which add cultural value to goods and make them more attractive to today's consumers. The design industries are predominantly small and micro

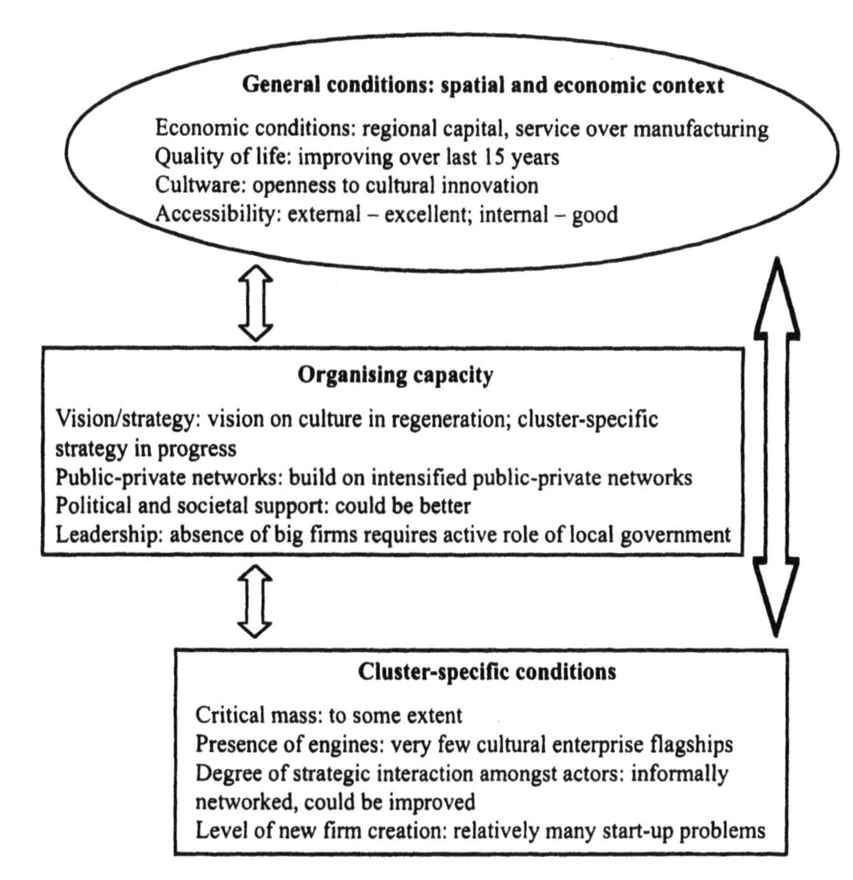

Figure 8.8 Confrontation with the research framework

enterprises and part of the supply chains of many other industries, which makes the sector 'invisible', one of its major problems. This conclusion also applies to most of the other subsectors in the cultural industries, excluding the media sector. There are not too many flagships (for instance big firms) either. In the media sector Granada TV is the flagship, but the traditional media sector is not expected to increase its economic spin-off in the Greater Manchester area. The BBC has cut down its operations in the area in favour of London. Although Granada TV is closely associated with the region, they too feel the gravitational pull of the London area. For most of the other subsectors the lack of flagship companies is a handicap.

There is, however, significant activity in the cultural industries in Manchester and in the field of design in particular, supported by local demand. The sector has critical mass to some extent, although starting up businesses in that field is still difficult, making the cluster vulnerable. To develop a good idea into goods and services generally takes a lot of time and effort. In a more economic vocabulary: these industries need to invest a lot of their very scarce resources in research and development. Venture capital, too, is still a problem for the starting up or expanding of cultural enterprise, although there are differences amongst the subsectors. In general, financial institutions know too little of the cultural business.

Paradoxically, some of the firms that overcome the start-up problems have managed to connect to interesting national, European and even global markets. Connection to the international markets is in many cases necessary because of the specialisation. The market for design is more developed in other countries than Britain where manufacturing industries are somewhat more reluctant to incorporate creative services and goods into their production. The linkages between the design, and the cultural businesses in general, and the education institutes could be improved. A better linkage could help to stem the leakage of creative talent to the London area.

An interesting phenomenon is that some of the more mature design businesses tend to cluster in the Castlefield area, while the Northern Quarter is preferred by activities that could be described as 'grassroots' and 'cutting edge'.

6.3 Organising Capacity

The organising capacity concerning the cluster revolves around four questions. First, has an integral vision and strategy been drawn up for the cluster? Second, do the public and private actors cooperate structurally? Third, has there been enough political and societal support? Fourth, is there sufficient leadership?

There is a vision for the role of culture for the Manchester area in the City Pride prospectus. The cultural industries are seen as a source of added value for commerce, industry and tourism. The City Pride vision has been translated into an arts and culture strategy for the City Pride area, in which the social, educational and economic aspects of arts and culture have a place. Manchester has adopted favourable policies (helped by European Funding and the Lottery Funds) for the arts and culture in general, but there are as yet very few policies in place for the cultural industries cluster itself. Nevertheless, the city has worked on a strategy addressing the specific needs of the cultural industries including tailor-made business support measures, and the city has opened the dialogue with the educational institutions on improving the match between the educational programmes and cultural enterprise.

As well as the city's bid for the Olympic Games and the successful bid for the Commonwealth Games, the City Pride initiative has boosted and structured public-private cooperation within the region. These joint efforts have made a significant difference in the relationship and communication between public and private actors in general. Representatives from Manchester's cultural and design community have been active in the bid process. The development concerning the criticised campaign 'We're up and going' of Marketing Manchester illustrates that in the field of the cultural industries cluster the communication between enterprise and the public sector could be improved. The political and societal support for the cultural industries could be improved too, but that has much to do with the lack of visibility of the sector.

When all is said and done, we have the impression that leadership in the cluster is still underdeveloped. In many other clusters in this comparative investigation, the larger firms or the ones with the international contacts orchestrate the internal cooperation. Due to the absence of these large enterprises in the field of design in the Manchester cluster, there could be a more active role for the public sector.

7 Conclusions and Perspectives

7.1 Conclusions

Research has confirmed that employment in the cultural industries is significant and that it holds promises for future growth. Manchester has the largest regional cultural industries sector in England. The largest subsector is media; however, projections for growth in this sector are modest in view of the gradual pull towards the capital, London. The creativity of Manchester's cultural sector does provide opportunities for future development of cultural enterprise in other subsectors such as design, both in the 'mainstream' and in activities 'on the cutting edge'.

Manchester's reputation in popular youth culture is a fruitful base for cultural enterprise to grow. However, for the cluster, and the field of design in particular, to develop further, its strength needs to be enhanced. Important areas of attention are the degree of economic organisation, the lack of visibility, the relation between enterprise and education, the special problems of starting-up cultural businesses (venture capital, entrepreneurial skills). Another point is the need to make more of the connections some designer firms have with international markets.

Manchester has acknowledged its cultural enterprise potential: now is the time to fulfil these ambitions. It is very significant that the vision in the City Pride manifesto for the city has been translated into a strategy and that concrete proposals are in progress. For these to be successful it is essential for the strategy to be given strong political support.

7.2 Perspectives

One of the main areas for improvement is the relation between cultural enterprise and the educational institutes. The design and cultural industries in general are a people's business in which investment in human capital is of the essence. Indeed, the student population of Manchester is a great economic asset. Manchester has one of the largest university campuses in Europe. It is an attractive place for students, largely due to the popular youth culture. Many talented young people from the North West Region and beyond come to Manchester to study. Twenty years ago, many Manchester students would leave the city after graduation to find a job in the Greater London area. Nowadays, the job opportunities have increased and a growing number of students stay in Manchester after they have finished their education. The number and sustainability of cultural enterprises could be raised if the training programmes of these vital human resources were more geared to this area of opportunity in Manchester. There is a role for the public sector to advocate such commitment to the various universities and institutes for further education.

Another main challenge is to increase visibility of cultural enterprise as an economic force in the city. Local demand provides a growth market for the design sector; however, access to global markets is also required. There is potential to increase the regional market if the availability of the services of these firms could be more effectively promoted. Cultural institutions such as Cornerhouse or the City Art Gallery or other large or small institutions (galleries) could play the role of flagships of cultural enterprise. CUBE, the newly-opened centre for urban design, has helped to raise the profile of design in Manchester and could be a vehicle for further promotion.

Manchester should stimulate local enterprise to use the international networks of cultural enterprises that have been successful in foreign markets. These firms are not very large, but the general feeling is that Manchester could make more out their successes. It is important to gain more insight into the state-of-the-art concerning the internationalisation of these businesses. How solidly are they rooted in the region? Could their ties with other firms in the region be reinforced? International successes could be used for the marketing of cultural enterprise both within the region and to the outside world.

There is a need for tailor-made policies for this type of enterprise. Manchester has worked to create the conditions for culture to contribute to the regeneration of Manchester in general. The field of cultural enterprise requires tailor-made policies to meet the need of new businesses, acquiring business skills. Manchester TEC has recently begun to respond more to the needs of media businesses. The measures outlined in the City Pride Priority Three Action Plan Delivery Strategy are important steps in the right direction, for which broad political support is needed.

There is a need for better coordination within the field of cultural enterprise. The degree of organisation in the sector is relatively low, but many informal networks exist. It is important for the sector to have its own voice to bring, as Northern Quarter Association has done, the problems and opportunities of the neighbourhood to a broader audience. There is a need for a new coordinating – a clearing house – or networking agency, after the example of the New York New Media Association. This has two major tasks: general marketing of new media enterprise and creating the platform for people working in the sector to meet on a regular basis. The proposed Cultural Industries Development Service could provide such a resource and visible focus for the sector.

Notes

1 The Manchester TEC area includes Manchester, Salford, Trafford and Tameside. The population in the TEC area is estimated to be just over 1.1 million people.
2 This is the so-called Labour Force Survey (LFS). This survey uses a wider International Labour Office (ILO) definition of unemployment. The ILO definition is different to the benefit claimant figures because it considers people unemployed if they have been actively seeking work in the four weeks prior to interview.
3 Bands like 'The Smiths' and 'New Order' gained wide international recognition.

References

Carter, D. (1997), *Creative Cities and the Information Society*, Manchester City Council.
City Pride Partnership (1994), *City Pride: A focus for the future.*
City Pride Partnership (1997), *City Pride 2 :Partnerships for a successful future.*
EDURC (Eurocities Economic Development and Regeneration Committee) (1997), *Growth Sectors in European Cities.*

European Commission (1998), *Culture, the Cultural Industries and Employment*, Commission Staff Working Paper (http://europa.eu.int/en/comm/dg10/culture/emploi-culture-intro_en.html).

Manchester Metropolitan University (1998), *The Prospectus*.

Manchester TEC (1998), *Economic Assessment 1998 – Part 3: Business and the Economy & part 4: Sectors*.

MIPC & DCA (1998), 'Cultural Production Strategy for Manchester', draft, May.

MTMC (1999), Manchester Multimedia Centre and Network Website, http://www.mmc.mmu.ac.uk.

Parkinson, M. (1998), *The United Kingdom*, in Berg, L. van den, E. Braun and J. van der Meer (eds), *National Urban Policies in the European Union*, Ashgate, Aldershot.

Russo, A.P. (1998), *The Cultural Sector and the Sustainability of Urban Tourism*, Erasmus University Rotterdam.

Discussion Partners

Mr G. Allman, Spoken Image Limited, Managing Director.
Mr P. Bailey, Cornerhouse, Exhibitions Director
Mrs L. Barbour, Manchester City Council, Arts and Culture Policy Officer.
Mrs J. Burns, Burns Owens Partnership.
Mr D. Carter, Manchester City Council, Principal Economic Initiatives Officer.
Mr J. Chapman, James Chapman Associates, Director.
Mr N. Clark, Manchester Multimedia Centre and Network, Director.
Mr J. O'Connor, Manchester Institute for Popular Culture, Director.
Mr L. Curtin, artist/designer.
Mr P. Griffin, independent film television producer .
Mr B. Harrison, Councillor, Chair of the Board of Manchester Airport.
Mr N. Johnson, Johnson UDC.
Mr R. Jones, Manchester Investment & Development Agency Service, Director, Marketing Operations.
Mrs A. McEvoy, North West Arts Board, Director, Visual Arts.
Mr D. Moutry, Cornerhouse, Director.
Mrs S. Rowlands, Northern Quarter Association.
Mr D. Sagar, architect.
Mr T. Speake, Manchester Institute of Telematics and Employment Research, Director.
Mr M. Starling, The Manchester School of Architecture, Director.
Mr M. Taylor, General Manager, Granada Television.

Chapter Nine

The Media Cluster in Munich

1 Introduction

This case study is dedicated to the growth cluster of media in Munich. Because the growth in the media sector is expected to come from new combinations of media and ICTs (information and communication technologies), the analysis will focus on the intersections of both sectors and on spatial-economic aspects of the growth processes.

The first four sections are predominantly descriptive; in the last section the findings will be put into the theoretical framework. Section 2 contains a brief overview of the development and structure of the Munich economy. Next, more specific attention is paid to the subsectors that can be discerned in the media sector: printing/publishing (section 3) and the audiovisual sector (section 4). For both subsectors, the analysis will focus on the formation of networks and the role of public policy. In section 5, the information and telecommunications sector is briefly depicted. Section 6 is dedicated to the new combinations that emerge at the crossings of media (both printing/publishing and the audiovisual sector) and information and telecommunications industries. Here also, the role of public policy in the emergence of new combinations will be described. In section 7, the media/ICT cluster will be put in the context of the analytical framework. Section 8 concludes.

2 Munich: Profile and Economic Development

2.1 Profile

Munich is located in the south of Germany. It is the capital of the Free State of Bavaria. It is the third largest city of Germany, after Berlin and Hamburg, with a population of about 1,300,000. Munich's population grew very rapidly after World War II. From the early 1970s on, population growth slowed down and even became negative in the early 1980s. From 1987 on the city's population has been growing once more, primarily owing to immigration. Unlike Munich itself, its surroundings have grown tremendously during the last 20 years. Since the 1960s the share of the 'Umland' in the regional population increased from around 37 per cent to 44 per cent in the early 1980s and to almost 48 per cent in 1993. Munich forms the centre of an urban region of 2,400,000 inhabitants.

Apart from the city of Munich, the metropolitan area of Munich consists of eight counties (Landkreisen) surrounding the city, varying in size from 97,000 to 278,000. These counties, in turn, comprise numerous relatively small but powerful municipalities. As a consequence of the rapid growth of the counties, commuter traffic has risen by 140 per cent since 1970. The reason for the heavy commuter flows is the concentration of jobs in the city of Munich and the spread of population across a much larger region. Two out of every three jobs are still in the city of Munich, but Munich hosts only half the population of the urban region.

2.2 Economic Structure and Development

The economy of both the city of Munich and the urban area is very well developed. The Munich region is known for its high living standards and its solid economic performance. It is counted amongst the most successful regions in Europe. From the 1980s to the early 1990s, the Munich economy grew almost constantly at a faster pace than the German economy. The Gross Regional Product of Munich doubled between 1980 and 1990. After 1992, the economy of Munich was hit by a recession.

Unemployment rates in Munich are very low. In 1998, only 5.8 percent of the Munich workforce was registered as unemployed, a remarkably low figure compared to other large German cities. Major German cities such as Hamburg, Cologne, Berlin and Düsseldorf all show unemployment figures well above 10 per cent.

The economic structure of the city of Munich and the urban region of Munich (often referred to as Planungsregion 14) is very well balanced, as is illustrated in Table 9.1. The distribution of activities amongst sectors does not differ substantially between city and surroundings, although service activities show a clear preference for the city. Munich's modern and balanced economic structure is often referred to as the 'Münchener Mischung' (Munich Mix). The Munich economy is balanced not only regarding the distribution of activities amongst sectors, but also with respect to firm size. Munich counts very large companies, such as headquarters of Siemens and BMW, but also numerous and very vital small and medium sized firms. In particular, Munich counts many innovative private firms and research institutions.

Table 9.1 Employees registered in social security system by sector in the Munich region, city and surrounding area, 1997

Sector	Region	Munich	Surrounding area
Agriculture	6,147	2,120	4,027
Energy and water supply	9,434	7,454	1,980
Manufacturing	229,436	139,843	89,593
Building, construction	49,396	30,058	19,338
Retail, wholesale trade	148,485	91,708	56,777
Transport, communications	58,599	36,566	22,033
Banking, insurance	72,420	63,379	9,041
Other services	331,276	221,095	110,181
Non-profit organisations, private households	33,903	28,311	5,592
Regional public authorities, social security	54,301	33,670	20,631
Secondary sector	29.0%	22.5%	32.1%
Tertiary sector	70.4%	72.6%	66.7%
Total	993,397	654,204	339,195

Munich's economic strength relates to its well-educated labour force. The education level of the Munich population is one of the highest among German cities. Fourteen per cent of the Munich working population has an academic degree, compared to 9.0 per cent in Cologne, 6.2 per cent in Dortmund and 6.0 per cent in Germany as a whole. The presence of such a well-educated labour force makes Munich a very attractive location for firms. Furthermore, Munich holds a very strong position in education. The city counts numerous institutions for higher education and has about 100,000 students (see Table 9.2).

Table 9.2 Number of students in Munich, 1997

University	Number	University	Number
Ludwig-Maximilians University	59,804	College of Political Sciences	577
Technical University	17,701	Conservatory	789
State Polytechnic University	14,306	Academy for Fine Arts	659
University of the Bundeswehr	2,158	College of Philosophy	383
Foundation Polytechnic University	1,509	College of Television and Film	316

Source: Referat für Arbeit und Wirtschaft, Landeshauptstadt München, 1996.

2.3 Media in Munich: a Growth Sector

Media is one of the most important sectors of the economy of the Munich urban region. In 1995, 6,700 firms were active in Munich and its surroundings in the media sector, generating a turnover of over DM 25 billion. Munich has many publishers, especially of technical and business magazines. The Munich region also holds a strong position in radio, TV and video. Statistically, the media sector in the Munich region is often divided into two subsectors: printing and publishing (print media, publishers, wholesalers), and audiovisual activities (film, photo, TV, radio, video). Strongly linked to the media sector is the marketing and communications sector (marketing agencies, congress organisers, market research).

In Table 9.3, the main figures about the two media sectors and the marketing branch are presented. From Table 9.3 it is very clear that media is a growth sector, in terms of number of firms, turnover and employment, although there are differences among the subsectors. In the 1988–95 period, the number of firms in the media and communication sector in Munich grew by 60 per cent. Turnover increased by 47 per cent. In printing and in the audiovisual sector, turnover grew twice as fast as employment. In the marketing sector, the reverse is true: turnover growth was very low (9 per cent) while employment grew fast (+26 per cent). Investments are by far highest in the audiovisual sector. From this table, the conclusion can be drawn that the media sector in Munich is booming, in terms of turnover, employment growth and investments. It is the fastest growing sector in the Munich economy.

Table 9.3 Some statistics on the media sector in Munich

	Printing/ publishing	Audiovisual activities	Marketing
Number of firms	2,074	1,545	3,078
Growth of the number of firms 1988–95	+36%	+61%	+83%
Fixed employment	38,195	20,032	13,833
Flexible employment	6,653	12,417	9,629
Employment growth 1988–95	+26%	+31%	+26%
Turnover (in billion DM)	11.3	8.9	4.9
Turnover growth 1988–96	+62%	+59%	+9%
Investments 1994 (in billion DM)	0.9	2.8	0.6

Source: Landeshauptstadt München, 1996.

In the next sections, we will deal with the printing/publishing and the audiovisual sectors separately, for two reasons. First, the activities in the two sectors are quite different and, second, partly as a consequence, we have the impression that network structures are scarcer between these subsectors than within them.

3 The Printing/Publishing Sector in Munich

3.1 Actors in the Sector

Munich counts 319 publishers. Every year, 14,000 publications roll off Munich's presses. Although, even before World War II, Munich played a considerable role in printing and publishing, it developed rapidly after the war, when many firms and people who had been active in publishing in Berlin and Leipzig came to Munich when the eastern part of Germany was occupied by the Soviets. Furthermore, Munich is very strong in newspaper publishing. Every day, five daily newspapers appear, with total daily sales of 1.2 million copies. In the magazine market, Munich, together with Hamburg, is the number one in Germany when the number of titles, editing boards and publishers are counted. One well-known title is the very successful *Focus* magazine, published by Burda Publishers. Technical, business and computer magazines are a particular strength of Munich. No fewer than 128 magazines (excluding trade journals) have their editorial offices in Munich.

As well as numerous publishers, Munich also hosts (vocational) training facilities related to the printing and publishing sector. The German College of Journalism is located in Munich, as are the Academy of German Booksellers and Publishers and the Vocational Training Centre for Printing, Design and Photography. Several congresses on publishing and books are held in Munich each year. Examples are the International Spring Book Week, and the Munich Media Congress.

3.2 Networks

Networks play an important role in Munich's publishing sector. There are many formal and informal relations between the firms and people active in the field. Authors, journalists, translators, publishers and wholesalers have frequent contact with each other and benefit from each other's services. Publishers, printers and other related firms in Munich benefit from the presence of a pool of well-educated people in the sector. Job mobility within the cluster is relatively high.

The large number of publishers and related activities, in combination with the networked character of the field, has contributed to the establishment of a joint facility called the 'Literaturhaus'. The Literaturhaus is located in the centre of Munich. Many institutes related to the field are located there: the Institute for Copyright and Media Law, the German Book Archive, the Society of Bavarian Publishers and Booksellers and the Lyrics Cabinet. The Literaturhaus functions as a formal and informal meeting place for the sector: it houses a 'trendy' grand café as well as exposition space, offices and meeting rooms. Lectures and seminars are held frequently, to inform actors in the sector of new developments relevant for the publishing field, for example on new media, IT and business aspects of publishing.

The Literaturhaus can be said to give the publishing sector in Munich a 'physical face' and stimulates the diffusion of innovations and the formation of networks. The concept is not entirely new. In other German cities with a strong publishing sector, like Berlin, Hamburg and Cologne, similar places exist. The typical aspect of the Munich Literaturhaus is the strong business orientation. In the programmes of the Literaturhaus, relatively much attention is paid to legal, business and technology aspects of publishing.

3.3 Public Policy

Both the City of Munich and the private sector are convinced of the advantages of the Literaturhaus as a meeting place, to facilitate networks, to stimulate the quick dispersal of innovations and information and to make the powerful publishing sector in Munich more visible. The financial contribution of both public and private sector illustrates this: the Literaturhaus is financed by the City of Munich (DM 20 million) and the private sector (DM 8 million).

4 The Audiovisual Sector

4.1 Actors in the Sector

The audiovisual sector (radio, TV, video and film) is strongly represented in the Munich region. In particular, there is a large presence of private radio and TV broadcasters. There are national broadcasters (Pro 7, DSF, Kabel 1, RTL 2, TM3), Bavarian-wide broadcasters (Sat 1, Bayern Aktuell, Bayern Journal) and local services (TV München, M1, RTL München Live). The most important public broadcaster is Bayerischer Rundfunk and the ZDF-Landesstudio. International broadcasters are also present, such as MTV and Eurosport. The importance of Munich as centre for private TV in Germany is overwhelming. Fifteen point two percent of the fixed employment of all private TV activities of Germany is located in the Munich area. However, if total employment in the audiovisual sector is regarded, Munich must cede the

German number one position to Cologne (see Table 9.4).

Table 9.4 Employment audiovisual activities (TV and radio) in the main German media cities, 1995

	Germany	Berlin	Hamburg	Cologne	Munich	Total cities' share
Fixed employment 1995	39,290	9.9 %	11.1 %	18.8 %	11.7 %	51.5 %
of which:						
public	30,278	10.0 %	9.9 %	20.9 %	10.6 %	51.3 %
private	9,012	9.7%	15.2 %	12.0 %	15.2 %	52.1 %
of which:						
television	22,402	10.1 %	12.0 %	17.8 %	13.1 %	53.1 %
radio	16,888	9.7 %	9.8 %	20.2 %	9.7 %	49.4 %
Other employment 1995	123,765	16.3 %	6.6 %	28.1 %	11.5 %	62.5 %
Total employment	163,055	14.8 %	7.7 %	25.9 %	11.5 %	59.9%

Source: DIW, 1997.

The strength of Munich becomes clear from Table 9.5: in terms of turnover in private broadcasting Munich is number one in Germany, with a share of 33 per cent of total German turnover. The strong presence of private broadcasting is partly due to the fact that Leo Kirch, a former film trader but now one of the most important players in the German private broadcasting scene, chose Munich as a location to start his commercial TV activities. Kirch's strong attachment and commitment to the region of Bavaria and the city of Munich largely decided his choice for Munich as a location. Note that most TV broadcasters are not located in the city of Munich itself, but in the neighbouring municipality of Unterföhrung.

Munich is a leading location in Germany for the film industry and even has a strong position on the international level tegarding film technology (cameras, light-technical, camera support). The most important players in the film industry in and around Munich are Bavaria (film technology), ARRI (camera producer) the Bavaria Studio GMBH (studio facilities) and the Kirch group.

4.2 Networks

The network relations in the audiovisual sector in Munich are of different

Table 9.5 Private broadcasting: turnover in millions of DM and shares of major cities, 1993

	Turnover	Share
Munich	3,000	33 %
Cologne	2,250	25 %
Mainz	1,290	14 %
Hamburg	850	9 %
Berlin	700	8 %
Other cities	1,010	11 %

Source: BLM, 1995.

types. First, there are many links between the broadcasters and various types of suppliers. DIW (1997) calculated that for Germany as a whole, the largest suppliers of TV stations are production firms (DM 4 billion). But broadcasters also spend much money on telecommunications (DM 1.2 billion) and agency provisions (DM 0.5 billion). The Munich-based TV broadcasters cooperate with suppliers of technology, artist agencies, marketing and communication firms and TV production firms, most of them located in Munich. Many TV productions are outsourced to firms in Munich, although TV broadcasters tend to produce more by themselves. The city is a good 'Standort' for productions, mainly because of the excellent infrastructure. However, in TV productions, Munich must leave the number one position to Cologne, the main reason being that, in Cologne, the local and regional government started very early to create a friendly climate for the media industry.

Many of the suppliers to the broadcasters are thus located in the Munich area. Surprisingly, however, one important element is missing. Despite the heavy local demand for TV commercials from the broadcasters, Munich on the whole lacks creative firms that are active in the development of these commercials. The TV stations buy such services outside Munich in other German cities (Frankfurt, Hamburg and Düsseldorf are German leaders in these creative activities), or even abroad.

In addition, there are networks amongst the broadcasters themselves. In the first place, they gain efficiency by making joint use of a studio complex, Bavaria Studio GmbH, just outside Munich. But the private broadcasters in Munich also participate in the Bavaria Academy for TV, thereby educating their future workforce. The academy was founded because the TV industry felt the need for practically-educated journalists and technicians. The courses of the existing film academy were considered insufficient for the needs of the TV stations. The teachers at the TV academy are mainly professionals from the TV broadcasters in Munich. Although students are free to find a job where they want, the interaction between students and teachers results in some 75 per cent of the students from the TV academy actually finding work in the Munich TV industry.

Finally, there are networks with large firms in the region. The presence of large firms in the Munich region can be said to play a positive role in the development of the TV-production sector. One example is the trend that large firms such as Siemens and BMW operate their own internal TV channels, or use video and audiovisual presentations. Production is often outsourced to the media producers in the Munich region.

4.3 Public Policy

In Germany, the regions (Länder) are for a large part responsible for media policies. Their tasks are, among other things, to regulate the distribution of broadcasting frequencies and to safeguard the quality of programmes and protect children. Apart from these prescribed tasks, however, the Free State of Bavaria pursues active policies to stimulate the development of media activities in Munich, often in cooperation with the private sector. First, it participates 40 per cent in the TV academy in Munich, and was even its initiator. Furthermore, Bavaria has initiated and co-financed a film and TV fund, the FilmFernsehFund (FFF). The FFF was founded in 1996 and serves as a 'one-stop shop', replacing all the scattered film supporting measures. The FFF has a budget of DM 50 million, the largest in Germany. The Free State contributes a yearly DM 27 million (56 per cent). Private investors, mostly broadcasters, raise the remaining DM 23 million. The aim of the fund is to stimulate the production of (TV) movies in the state of Bavaria. Producers who make use of the fund are obliged to record certain parts of the movie in the Free State. The fund should contribute to the image of Bavaria as a movie region. The city of Munich can also benefit from it, for most of the producers, post-production firms, film-technical firms, actors and others are located in Munich. Since the foundation of the FFF, the number of supported films recorded in Bavaria has grown substantially (Blickpunkt Film, 1997).

Second, the Free State of Bavaria operates an organisation for media, the agency for new media. This agency is very active in monitoring the new media industry in Bavaria (collecting and publishing of statistics and trends, conducting research). In addition, it is involved in the acquisition of new media firms to the Bavaria region, and the support of new companies in the sector. The organisation is furthermore active in supporting media-related education: it contributes, for example, to a media marketing course at the Media Academy in Munich.

In contrast to the entrepreneurial attitude of the Free State of Bavaria towards the media, the role of the municipality of Munich in the audiovisual sector was very limited in the boom times until 1992. The political climate and attitude in the city of Munich in the 1980s towards media, in particular private broadcasters, was restrained. It may well be that this contributed to the location of private TV broadcasters just outside the city of Munich, in the municipality of Unterföhring, in the 1980s.

In recent years however, the attitude of Munich towards new media has changed: the Municipality has increased economic policy efforts due to new (economic) developments. Considering the economic importance of the media, Munich is pursuing now a positive, partly initiating role: it developed an action plan to develop Munich further as centre of multimedia and telecommunication, with various pronounced activities:

- promotion of events like the 'Münchner Filmfest' (Munich Film Festival), the Media-Days Munich, New Media awards;
- support of the organisation of media fairs;
- funding of the very prestigious Corporate Video & TV – New Media competition since 1998;
- installing the 'Munich Multimedia Academy' for new professions in the field of multimedia;
- fostering vocational training in the field of multimedia, i.e. by an EU-ADAPT project, the 'Gesellschaft für Medienberufe';
- furthering of 'Media Works Munich', an emerging complex in the city for the new media industries in synergy with media industry institutions;
- planning a new multimedia centre to widen civic access;
- city marketing: development of an image plan for fostering Munich as a city of new media.

However, the contrast with Cologne, another big new media city in Germany, has been considerable: Cologne, together with the region of North Rhine-Westfalia, has pursued an active, stimulating policy towards new media since the beginning of the era of private TV, as a means to revitalise the outdated industrial base of the city. Firms in the media field were offered 'soft' loans and facilities. It is generally recognised that the early efforts and policies of both the City of Cologne and the region of North Rhine-Westfalia have contributed to the flourishing of the new media in Cologne.

5 Telecommunications and Information Technology

5.1 Actors in the Sector

As pointed out, Munich is a city with a rich diversity of media firms. But Munich also holds a strong position in the field of telematics, telecommunications and software. The position of Munich as location for media and telematics has been analysed in an investigation by Landeshauptstadt München (1996). Munich proves to be very strong in electronic data management and telecommunications industry (including software). Apart from Siemens, 22,000 people work in these fields. In particular, there is a strong presence of computer and software companies from the USA. Most of them are active in sales, service and education, and not so much in production. In electronic data management and the telecommunications industry (including software), Munich is number one in Germany, before Berlin, Hamburg and Cologne. Small and medium-sized firms dominate the scene, but there is also a number of big firms.

5.2 Major Players

There are two main providers of telecommunication services present in the city: a regional headquarters of the former state monopoly Deutsche Telekom and the national headquarters of the recently founded new telecommunications provider, VIAG-Interkom. VIAG-Interkom has been one of the main competitors of Deutsche Telekom since the liberalisation of the telecommunications market in Germany of 1 January 1998.

The former state monopolist Deutsche Telekom (DT) has long been absent from the field of new media. Recently however, a subsidiary company, T-media, was established. This organisation offers services in new media, such as Internet access and the development of services for cable TV and digital TV. Deutsche Telekom decided to locate T-media in Munich, mainly because of the good 'climate' in the city for this type of service. Because DT lacks critical technological know-how, it cooperates as much as possible with firms that do have the technological know-how to develop these new products and services. Even so, T-media experiences heavy competition from new firms that are active in these fields.

In contrast to Deutsche Telekom, which is a centrally organised firm directed from the headquarters in Bonn, the new telecommunications provider VIAG Interkom is strongly connected with the Free State of Bavaria and the

Table 9.6 Large firms in media and communication, 1995

	Hamburg	Berlin	Cologne	Munich
Number of large firms	76	74	29	80
Total number of employees	68,200	34,000	9,300	179,000
of which:				
telecom/software industry	10,400	14,900	3,200	22,800
telecom/software services	450	3,100	450	1,420
publishers	25,000	4,600	3,650	7,030
film/TV/video	900	640	910	920
others	2,400	1,100	1,200	1,400

Source: Prognos, 1997.

city of Munich. Forty-five per cent of the shares are in the hands of VIAG, the energy company of Bavaria. VIAG Interkom uses the cable networks of the mother company for data transmission. On the demand side the links with Bavaria are strong: VIAG Interkom provides many telecommunication services to connect offices of the Free State of Bavaria, and offers special arrangements for big firms in Munich, especially in the media sector. On the suppliers' side, VIAG Interkom makes heavy use of suppliers in Munich and Bavaria.

From 1997 to 2007, VIAG Intercom has planned to expand rapidly: the firm expects employment to grow by 9,000 jobs, 2,000 of which will be at the Munich headquarters. Such rapid growth will also be beneficial to suppliers, which are, for a large part, small high-tech firms in Munich and Bavaria.

5.3 Public Policies

Despite the strength of the sector, there are many initiatives – by both the Free State of Bavaria and the city of Munich – to stimulate the development of the information and communication technology sector even more. The city of Munich recently developed an action plan to develop Munich further as centre of multimedia and telecommunication. Part of the strategy is to enhance the electronic infrastructure and to stimulate the organisation of fairs in the field of telecommunication. In the field of education, the city seeks to ameliorate the existing infrastructure: it is active in the development of a multimedia academy, in helping schools to get online, and in stimulating vocational education in the field of multimedia. Further initiatives have been undertaken to use interactive media in the city administration and in its tourist policies (Landeshauptstadt München, 1998).

Particularly interesting for the dynamics of the cluster is the development of the MTZ (Münchner Technologie Zentrum, Munich Technology Centre) in the northern part of the city. The aim of this centre is to help promising aspiring entrepreneurs in new fields – multimedia and telecommunication amongst them – to increase their chances. Entrepreneurs can get cheap accommodation and are offered all kinds of advice (financial, legal and technical). Interestingly, much stress is laid on creating networks between young and well-established firms in the field of ICT, thus facilitating knowledge and technology transfer. Already, some 20 young firms in the fields of media/ICT are located in the centre. The initiative is strongly supported by the city of Munich, the Chamber of Commerce and the region of Upper Bavaria (http://www.mtz.de).

The Free State of Bavaria is also an important policy actor: it pursues active policies to stimulate the use of ICTs and new media in Bavaria. Its most important project is the Bavaria Online project. The aim of the project is to connect as many people, firms and organisations to the Internet as possible. Competition with private sector access providers is avoided. The project is clearly a public-private initiative: apart from the financial contribution of Bavaria, private firms invest DM 350 million. Further, the Bavaria Online project operates many pilot projects (see Table 9.7).

One of the pilot projects of Bavaria Online is teleworking. In this project, a telework programme has started with BMW. The aim of the project is to enable 400 employees of the Munich-based car producer BMW to work at home.

The Free State of Bavaria plays an important stimulating role in the development of new activities at the crossing of media and ICT. It is supporting future-oriented projects in new media, with investments of a total of DM 100 million. Regarding the audiovisual sector, the regional government stimulates the development of new broadcasting techniques. For example, it operates a pilot project aimed at the development of digital radio broadcasting. In this project, the Free State of Bavaria cooperates with Rohde & Schwarz GmbH, a large Munich-based firm specialising in transmission technology. The experiences in Bavaria with digital radio are very positive. As a result, the concept is now an export product of the region.

6 New Combinations

The very prominent presence of both media activities and information

Table 9.7 Pilot projects of Bavaria Online

Project field	Projects	Aim
Schools, education	Bavarian school-net Training-programme for teachers	Internet connection for all Bavarian schools; ICTs in education at school Training of teachers in ICTs, teleteaching
Public	Bavarian Civic Net	Stimulate the civic use of the Internet
Universities	University Net Multimedia at universities	High-speed networking of universities and research institutes Multimedia for university teachers
Safety	'Basilika' project for electronic data security	Security architecture for the 'Bayernnet'
Transport	Freight transport logistics 'Bavaria 2000'	Intelligent freight bundling and integrated freight management
Businesses	Information for SMEs Textile industry multimedia database City Net Munich and City Net Nuremberg	Internet network and data pools to stimulate the use of ICT in SMEs Stimulate fast online communication within the textile industry Stimulating high-speed fibre optic rings
Teleworking	Teleworking in a conurbation	Stimulate teleworking in large businesses
Medical	Health Net	Use of ICT in medical sector for teleconsultation
Technical	Digital Audio Broadcasting Bavarian Innovation Net	Development of digital radio Close regional connection between research and development and the innovative industry
Government	Authorities Net 'Solumn Star Net' Bavaria Server	Install a corporate network for the Bavarian authorities Digital land register Internet and email server for information services of the Freestate

Source: Bavaria Online, 1998.

technology firms makes many synergies possible and provides for excellent dynamics. Moreover, the presence of local demand in the strong Munich economy boosts the new combinations. The emergence of new products and services at the intersections of media and telecommunications entails new value chains in the production and distribution of these products and services. Complex interfirm networks are created by progressive players in the field of media and information technology to benefit from the vast range of possibilities.

6.1 Audiovisual Activities and ICT

In the audiovisual sector there is a strong convergence of information technology and media activities. This tendency has two aspects. First, existing, or 'traditional' activities are digitalised and second, the new information possibilities lead to new products and services. Examples of the digitalisation of 'traditional activities' in the media field are films and videos made with the aid of computer programs and digital techniques, resulting in a better quality of sound and vision. The same holds for TV productions.

Secondly, and more importantly, the new information technologies and infrastructures entail new products and services. There are many examples: pay-per-view concepts, interactive TV, information services for car drivers, music- and video-on-demand services. The core concept of these new products is individualisation: individual customers no longer buy a 'standard' product, but choose their own programmes. The new products are only at the beginning of their life cycle. For example, pay-per-view and video-on-demand services are not yet widely used and accepted. For the future, however, much growth can be expected in these products and services. Anticipating future growth opportunities, the private TV and radio broadcasters in Munich are very active in digital broadcasting. Recently, the Kirch Gruppe started the station DF1 (Digital Fernsehen). Broadcasters also try to capture new markets via the Internet. Private broadcaster Pro-7 for example operates an online publishing platform, through which all kinds of services are offered.

6.2 Printing/Publishing and ICT

In the printing and publishing sector, ICTs entail first a digitalisation of existing activities, second the development of new activities and third a different spatial distribution of activities. On the one hand, digital techniques replace 'old' technologies in publishing and printing: pre-print activities are executed with computers, digital techniques are used in the printing process, design is computerised. Publishers use new media such as CD-ROM to store information. On the other hand, the new information technologies and infrastructures lead to new products and services. Electronic publishing is expanding rapidly. Quality newspapers are published on the Internet (often with free access, revenues coming from advertisements). For tabloids, the Internet offers fewer opportunities: the product is less suited for online publication and the customers are generally less familiar with the new medium.

For publishers of business information, new electronic media offer vast prospects. Some publishers in Munich already offer subscriptions for specific business information on the Internet, adapted to the demands of the customer. Electronic publishing firms are growing rapidly. Most of them are subsidiaries of traditional publishers. Although Munich has a good position in electronic publishing, in Germany it lags behind the cities of Karlsruhe, Hamburg and Cologne.

Interestingly, the rapid development in ICTs enables publishing firms to choose a different spatial organisation. In many cases, publishing firms separate their editing activities and printing activities spatially. Many publishers (for example Süddeutsche Zeitung and Merkur) have their editorial room in the centre of Munich and the printing facilities outside the city, where land prices are generally lower. The editing and printing locations are connected by dedicated high-volume data transmission lines, established by tele-communication firms. Prognos (1997) listed the new products and services in the digital media as depicted in Figure 9.1.

7 Confrontation with the Framework

In the last few sections, we have tried to describe some of the processes at work in the growth sectors media and ICTs in the Munich area. In this section, the development in the growth sectors is placed the framework of reference presented in Chapter Two. First, we will discuss the urban context in which the media cluster is embedded. We argue that the position of the Munich economy in general is very favourable, at a time when economic development depends on networks and knowledge. Second, we turn tc the specific development of the growth sector of both media and IT/telecommunications activities in Munich. Third, we will discuss the organising capacity regarding the cluster.

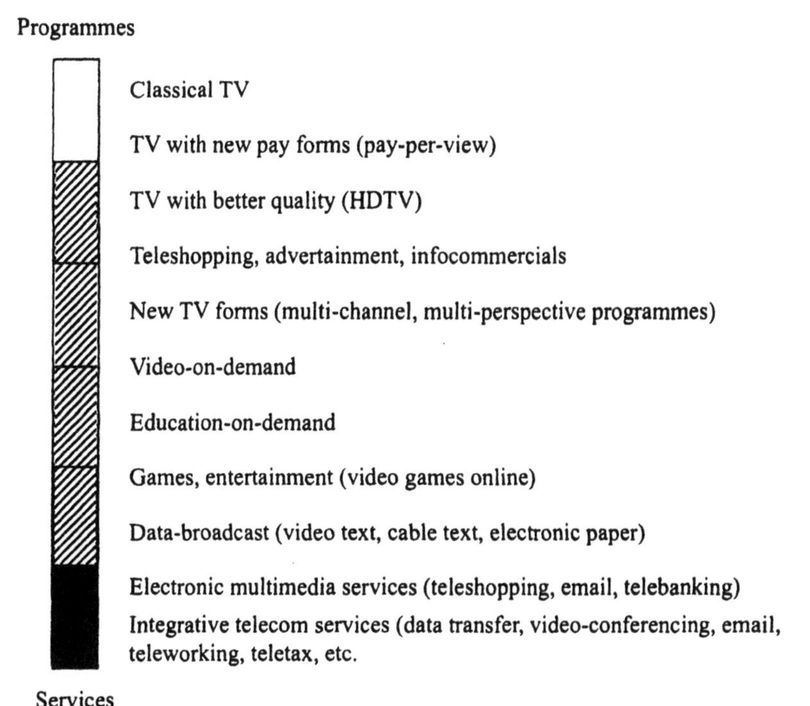

Figure 9.1 The spectrum of new audiovisual activities

7.1 General Conditions in Munich

The economic strength of Munich can be derived from its diversified and modern economic structure and the strength in different networks, both internal and external, as well as its strong local knowledge base. The corporate control centres, as well as the countless other high-grade firms in Munich, rely and depend on the well educated, highly-skilled people that are available in Munich. As the economy is so strong, the demand for media products in Munich is very well developed, which benefits the regional media cluster to a large degree.

The spatial conditions in Munich are on the whole favourable. First, Munich's external (international) accessibility is excellent. Apart from highway access, Munich has connections with many European other destinations, by air or by train. A location factor of increasing importance – for firms in general but for firms in the media and communication industry in particular – is the presence and quality of a high-quality telecommunications infrastructure. Munich proves to be quite strong on this point compared to other large German cities (Prognos, 1997): the internal and external 'electronic accessibility' of

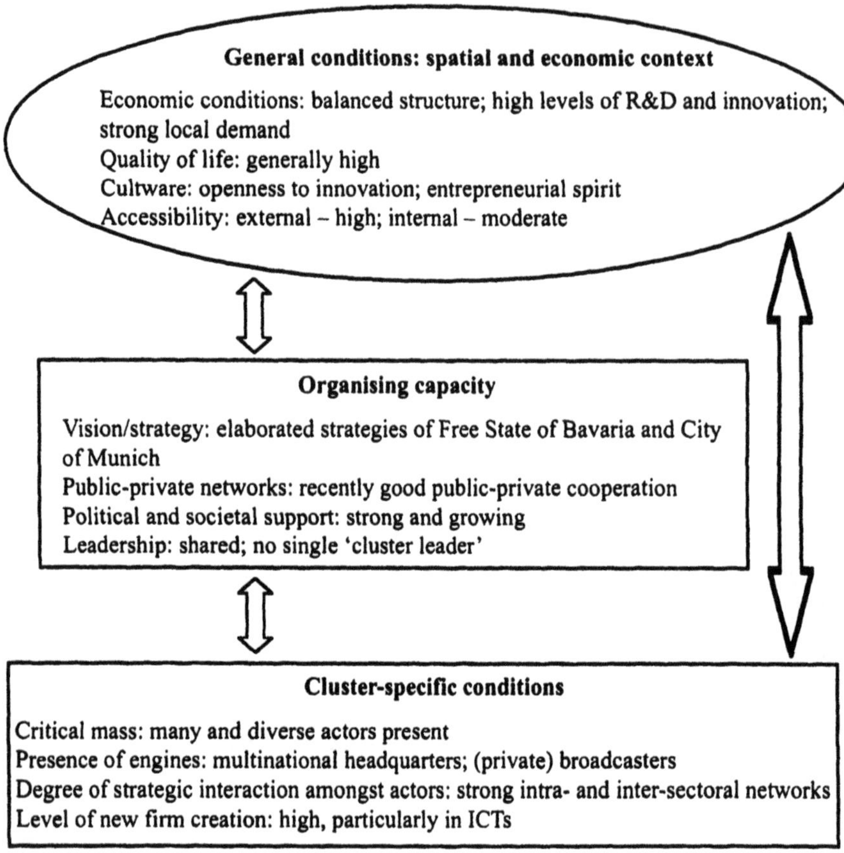

General conditions: spatial and economic context

Economic conditions: balanced structure; high levels of R&D and innovation; strong local demand
Quality of life: generally high
Cultware: openness to innovation; entrepreneurial spirit
Accessibility: external – high; internal – moderate

Organising capacity

Vision/strategy: elaborated strategies of Free State of Bavaria and City of Munich
Public-private networks: recently good public-private cooperation
Political and societal support: strong and growing
Leadership: shared; no single 'cluster leader'

Cluster-specific conditions

Critical mass: many and diverse actors present
Presence of engines: multinational headquarters; (private) broadcasters
Degree of strategic interaction amongst actors: strong intra- and inter-sectoral networks
Level of new firm creation: high, particularly in ICTs

Figure 9.2 Framework of reference

Munich is excellent. The networked structure of the Munich economy both internal and external is thus supported by its physical networks, although it should be noted that internal accessibility is moderate.

Intimately related to the strong knowledge base of Munich is the quality of life in the Munich area. The city of Munich is generally considered a pleasant place to live, although very expensive. Many high-quality cultural facilities (museums, orchestras, opera) are present, the housing stock is excellent and there are many luxury shops. Another aspect which makes Munich attractive is its proximity to the countryside. Skiing resorts in the Alps are very close, as are big lakes. The quality of life that Munich offers is the fundamental prerequisite for the knowledge base of the city: it makes Munich attractive to skilled people with high demands on the quality of life. It also helps expanding firms in media and communications to attract staff.

Another asset of Munich is its 'cultware': generally speaking, the attitude of people and firms in Munich can be characterised as 'open to innovation'. On the firm level, that openness undoubtedly facilitates knowledge and information exchanges, technology transfer and the formation of innovative cooperative networks amongst firms and knowledge institutes in the Munich area, thereby contributing to the high 'innovating capacity' of the region. The openness to innovation of the inhabitants of the region leads to a relatively fast adoption of new technologies by the public, such as digital broadcasting, electronic shopping etc. The openness to innovation in Munich is first and foremost related to the technical and economic spheres. Interestingly, the municipality invests in a further strengthening of the climate for new technologies and acceptance thereof by civilians, together with education institutes and firms.

7.2 Strength of the Media Cluster in Munich

It is in the above sketched context of the Munich economy that the media cluster should be studied. The networked character, the strong knowledge base and the innovative spirit, so characteristic of the Munich economy as a whole, can be found in the media sector as well, and contribute to new combinations and rapid development. As we have seen, the cluster has critical mass: many elements of the media cluster in Munich are present: publishers, printing facilities, marketing agencies, broadcasters, producers, technology suppliers and much more. Only one type of element is virtually absent in the Munich media cluster: the 'creative agents', firms that are involved in the creative genesis of commercials.

Very importantly, the elements do not function apart from one another, but are often strongly linked. There are numerous interfirm and interpersonal networks, very important for media firms because the networks are a means for the diffusion of new ideas and innovations. In Munich, an important aspect in the successful development of the media sector is that the network formation does not remain limited to firms within the media (sub)sectors: cross-fertilisation between sectors or intersectoral networks is also well developed. For example, are relations between media firms and telecommunications providers or between editors and Internet specialists are very important. These networks result in new combinations and are the vehicles of the continuous 'creative destructive' processes beneficial for media firms. The joint concentration of a large number of large and small telecommunications and software companies in Munich and the media sector together provide an enormous potential for the Munich economy. The possibilities are enhanced by the presence of large firms in the Munich regions, which often act as buyers of new innovative products.

The presence of critical mass in the media sector in Munich is an important 'enabling' aspect of the strength of Munich in the media sector. An advantage of the concentration of media firms is the possibility of sharing facilities and resources, or setting up projects. For example, the critical mass of the publishing firms in Munich enabled firms together with the municipality to establish the Literaturhaus, thereby further strengthening the sector. The concentration of broadcasters made the joint foundation of a TV academy possible.

7.3 Organising Capacity

The high degree of organising capacity, which can be defined as the capacity to enlist actors and develop and implement policies regarding the media cluster, is an important success factor in the media cluster in Munich. Both the city of Munich and the Free State of Bavaria clearly recognise the importance of media – in the broad sense – as an economic sector and pursue active strategies to make as much as possible out of it.

Regarding public-private networks, the Free State of Bavaria has a leading role. As becomes clear from Table 9.7, the Free State of Bavaria is able to enlist actors from the private sector on a large scale. Illustrations are the foundation of the TV academy by the Free State of Bavaria and the private sector, the setting up of a film fund, and measures taken by the Free State to develop new media and to speed up their adoption by firms and inhabitants. In the second place, the Free State of Bavaria acts as a critical consumer of new services, thereby stimulating innovation in the private sector. Examples of 'demanding projects' are the linking of networks of offices, and the development of digital broadcast technology.

The existence of so many joint projects and initiatives in different branches of the media sector and at the frontiers of the media sector and other sectors form an indication of the cooperativeness and the cluster character of the media industry in the Munich region.

In the boom times of commercial television, before 1992, the municipality of Munich was less active in the media sector. Regarding new media such as private television, the attitude of the municipality has been restrained for too long. This may have had consequences for the position of Munich as a location for private broadcasters, in particular compared to Cologne. The same holds

Table 9.8 Some public-private media-related projects in the Munich region

Product/service	Project/facility	Participants	Financial contribution in %
Traditional publishing	**Literaturhaus**	Publishers	29%
E-publishing		Municipality	71%
Private TV	**TV Academy**	Broadcasters	50%
Digital TV/video		Free State of Bavaria	40%
Digital radio		Private sector	70%
Teleworking	**Bavaria online**	Free State of Bavaria	30%
E-networks			
Films (cinema and TV)	**Film and TV Fund**	Broadcasters/producers	44%
		Free State of Bavaria	56%

for the attitude of the Municipality towards the film industry. In the publishing sector, the municipality recently played a positive role, in particular by contributing to the establishment of the Literaturhaus. In recent years, political (and societal) support for the media cluster development has been growing, a development that is reflected in the recent action plan for the development of Munich as centre for multimedia and telecommunication.

8 Conclusions and Perspectives

The media cluster in Munich is very strong, well developed, diversified and modern. The strength of the cluster can partly be ascribed to historic factors – the German divide after the Second World War made many media firms flee to Munich – but the strong economy of the city of Munich as a whole should also be regarded as important factor in the success of the cluster: many parts of the cluster – notably telecommunications, new media, marketing and communication activities – are strongly stimulated by the local demand. Another characteristic is the strong network culture in the cluster: strategic linkages prevail, facilitated by cultural identity and a cooperative attitude amongst actors. This leads to new combinations and permits cluster actors to react rapidly to new developments.

Although the cluster is clearly dominated by private business, the role of public bodies (notably the city of Munich and the Free State of Bavaria) in the stimulation of the cluster is considerable. The city of Munich, with its initial disregard of the new media sector, particularly private television, has now taken a positive and stimulating attitude towards the cluster, from the vision that the cluster has much economic impact and may increase the quality of the cities' economic structure. The Free State of Bavaria has a longer tradition in active media policy, in particular in new media stimulation, and contributes to cluster development as 'critical demander' of new ICT services.

The technological sophistication of the cluster is high, due to the presence of excellent, world-class firms and high-grade universities in the fields of ICT and new media. This gives Munich a lead over other media cities, in particular in the promising new combinations of 'traditional' media and information and telecommunication technology. It is expected that future growth will occur in these fields. Prognos estimates an employment growth of 8,000 in the period from 1995–2000, and another 5,000 by 2010. From this observation, it may be stated that the cluster will probably develop further in the (near) future. Another hopeful sign is the high level of new firm creation in the cluster, partly due to the success of the Technology Centre, but more generally to the favourable 'entrepreneurial spirit' in the city.

References

Bayerische Landeszentrale für Neue Medien (1995), *Hörfunk und Fernsehen als Wirtschaftsfaktor.*

Blickpunkt Film (1997), *Bayerischer Königsweg macht Schule, Blickpunkt Film 17–2–1997.*

Deutsches Institut für Wirtschaftsforschung (1997), *Beschäftigte und wirtschaftliche Lage des Rundfunks in Deutschland 1995.*

Faltlhauser, K.(1997), *Statement von Herrn Staatsminister Prof. Dr. Kurt Faltlhauser anlässlich der Pressekonferenz zum Beschluss des Bayerichen Ministerrats über das Förderkonzept BayernOnline II.*

Landeshauptstadt München, Referat für Arbeit und Wirtschaft (1996), *München, der Wirtschaftsstandort.*

Landeshauptstadt München, Referat für Arbeit und Wirtschaft (1996), *München, Medienhauptstadt Deutschlands.*

Landeshauptstadt München (1998), *Wirtschafts- und beschäftigungspolitisches Konzept für München*, December 1998.

Prognos (1997), *Standortbewertung für die Stadt München im Bereich Telematik und Medien.*

Discussion Partners

Mr Beckman, Siemens AG, Vice President and Director, Customer Marketing.
Mr H. Gasser, Zeitungsgruppe Münchener Merkur, General Manager.
Mr Kiemer, IFO Institute, researcher.
Mr Krauss, Deutsche Telekom, Marketing/Communication Business.

Mr Kors, Bayerische Landeszentrale für Neue Medien, Head of press affairs, Public Communication and Media Industry.

Mr Hansjörg Kuch, Staatskanzlei Freistaat Bayern, Department of Press, Public Communication and Media.

Mrs A. Seip, Communication and Press, MGM Mediengruppe München.

Mr Klaus Schussmann, City of Munich, Department for Labour and Economic Development.

Mr R. Simionescu, Bayerische Akademie für Fernsehen.

Mr Ulmer, VIAG Interkom.

Mr Witman, Literaturhaus.

Chapter Ten

The Audiovisual Cluster in Rotterdam

1 Introduction

In this case study, the audiovisual cluster in the Rotterdam urban region will be analysed. It is generally recognised that audiovisual production is a growth sector: the European Union recently reported that the audiovisual industry has a great potential for growth and job creation (EU, 1993). The rapid development of technology combined with the growing demand for communication from society and business both contribute to that potential. However, every city or urban region should find its own way to benefit as much as possible from the growth of this cluster. For Rotterdam, this is particularly relevant, for the city does not have a great tradition in media and communication compared to many other cities, such as Amsterdam in the Netherlands or Munich in Germany. In this study, we will put the audiovisual sector in the perspective of the Rotterdam economy as a whole. We will particularly stress the relations between audiovisual firms themselves, education facilities and large business as the main demanders or buyers of audiovisual products.

This chapter is structured as follows. Section 2 contains a brief description of the economic structure of the Rotterdam region. In section 3, we will describe the different elements of the media cluster in Rotterdam, comprising media firms/institutes, education and the 'demand side'. Section 4 is dedicated to the analysis of interaction between the elements. In section 5 the growth cluster of multimedia is put in the framework of reference of chapter 2. Finally, section 6 contains some conclusions and perspectives.

2 Rotterdam: Profile and Economic Development

Rotterdam is situated in the southern part of the province of South Holland, and is a part of Randstad. It is the second largest city of the Netherlands, after Amsterdam. The city of Rotterdam counts some 600,000 inhabitants. The Rijnmond Region, consisting of the city of Rotterdam and its surrounding municipalities, counts 1.1 million inhabitants.

With a share of 10 per cent of the GDP of the Netherlands, the Rijnmond region is one of the most important economic engines of the Netherlands. The large seaport contributes generously to this economic power. Many economic activities are directly or indirectly related to the port. Directly port related activities are transport, transhipment and logistics activities, and the petrochemical industry. Indirectly, the port is a driver behind service activities that support the port-related businesses. Examples are banking, insurance and all kinds of business services. The economic structure of the Rotterdam economy is shown in Table 10.1. Particularly strong are the transport and distribution sector, and the financial and business services sectors. During the last two decades, fundamental changes have affected the sectoral structure of the Rotterdam economy. As in many other European cities, the share of industrial activity has declined, and that of service activity grown substantially: see Figure 10.1.

Table 10.1 Economic structure of Rotterdam, 1996

Sector	People employed	Sector	People employed
Transport, distribution/communication	40,512	Agriculture	501
Financial services	16,590	Energy extraction	9
Business services	33,260	Industry	31,157
Public administration	12,806	Public utilities	1,853
Education	17,712	Construction	15,020
Health and welfare	35,360	Wholesale	15,523
Culture, recreation and other services	9,190	Retail	22,627
Hotel and catering industry	7,174	Total	260,421

Source: COS, 1996.

Within industry, important changes have occurred: industrial activities such as shipbuilding and food industry have lost their dominant position in the last 20 years. The decreasing share of port activities and port related activities in the economy is characteristic. Although the port has grown steadily in terms of transhipment and goods handling, employment in port activities has decreased (Van Klink, 1996).

Private enterprise in Rotterdam is dominated by small and medium sized enterprises, as can be seen in Table 10.2. By far the largest number of firms employ fewer than 50 people. The number of firms employing more than 100 employees is relatively small. However, these firms generate more than one half of the total employment in the Rotterdam region. The top 100 of Rotterdam

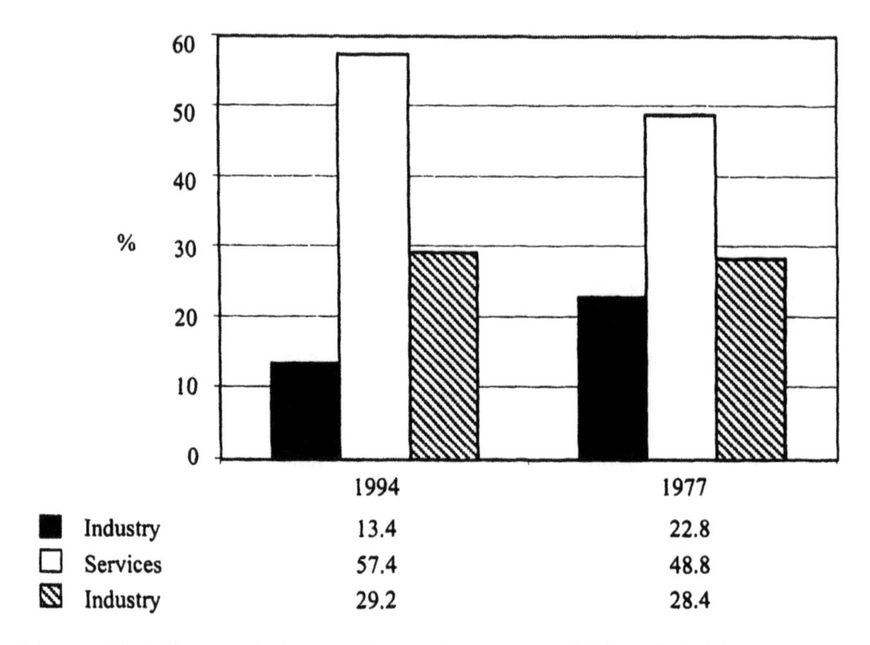

Figure 10.1 Sectoral change in employment, 1977 and 1994

Source: Chamber of Commerce, 1995.

Table 10.2 Firms in Rotterdam, sectors and size classes, 1996

Sector	0–49	50–99	>100	Total
Agriculture	171	–	–	171
Energy	4	–	–	4
Industry	1,080	39	63	1,182
Public utilities	2	2	8	12
Construction	1,049	35	22	1,106
Wholesale	1,914	38	19	1,971
Retail	5,266	20	9	5,295
Hotel and catering industry	1,718	10	5	1,733
Transport, distribution/communication	1,957	70	62	2,089
Financial services	672	21	27	720
Business services	3,469	69	59	3,597
Public administration	90	33	41	164
Education	718	51	21	790
Health and welfare	1,269	46	62	1,377
Culture, recreation and other services	1,654	13	12	1,679
Total	21,033	447	410	21,890

Source: COS, 1996.

firms (ranked by turnover) is dominated by firms in the petrochemical industry, trade activities and transport and distribution: see Figure 10.2.

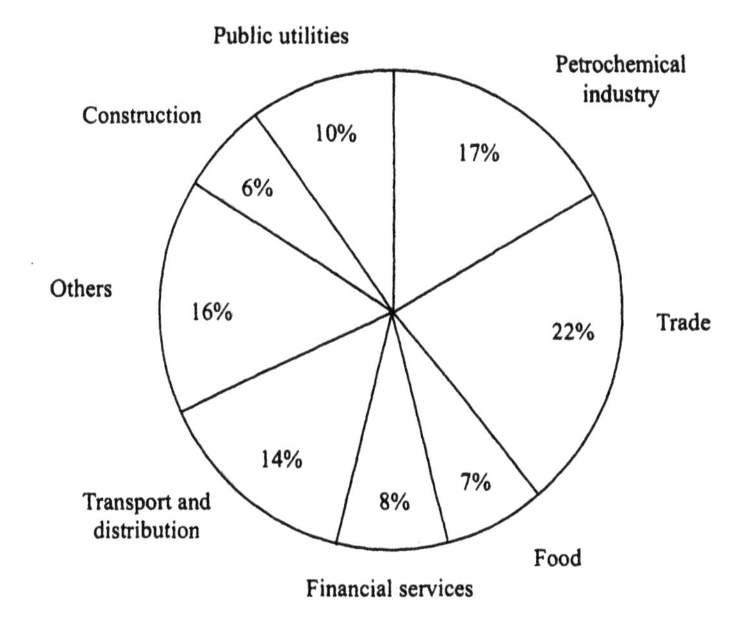

Figure 10.2 The 100 largest firms in Rotterdam: sector division

Source: KPMG, 1996.

A particular problem for Rotterdam is the fact that transport and distribution activities in the port, though still very important, generate fewer and fewer jobs in the region. The economic benefits of the expansion of the port of Rotterdam accrue mainly to inland regions, where transport and distribution activities increasingly locate (Van Klink, 1996). With this in mind, the economic strategy of the municipality is directed to the broadening of the economic base of the city. In particular, the municipality focuses on a few sectors that are regarded as potential generators of economic and employment growth. One of these sectors is the audiovisual industry; others are tourism, medical technology and the recycling industry.

3 The Audiovisual Cluster in Rotterdam

In this section, we will describe the audiovisual cluster in Rotterdam. In line with the cluster concept developed in Chapter One, we will consider not only the private firms that belong to the audiovisual sector, but also other important players in the audiovisual sphere, such as educational institutes and government bodies. Furthermore, we will pay attention to the large customers of audiovisual services in the region: broadcasters and large firms. The analysis of the interaction between the cluster elements is the subject of section 4. For a description of developments and trends in the media sector in general, we refer to Chapter Six, section 3.

The audiovisual industry in the Netherlands
Statistics on the size of the audiovisual industry in the Netherlands are scarce. McKinsey (1993) estimates that the total production value of the audiovisual industry in the Netherlands amounts to some Dfl 1.1 billion, of which TV is over 60 per cent and film 4 per cent. The remainder includes business-to-business communications (28 per cent) and commercials (7 per cent). Compared to other small countries, the Dutch TV production industry is relatively strong in economic value. The main reason is the presence of a few strong TV production companies that export programmes and programme concepts on a large scale. One of the production firms, Endemol, is even one of Europe's largest production companies. After the UK and Italy, the Netherlands is the third largest exporter of TV programmes in Europe (McKinsey, 1993). The great Dutch facility houses, in particular the NOB, operate relatively cost-efficiently and are technically strong. In the Netherlands, film production is weakly developed. The domestic audience share of Dutch feature films had dropped to some 5.4 per cent in 1996 (Wolff, 1998), one of the lowest shares in Europe. The audiovisual industry in the Netherlands is concentrated in the area between Amsterdam and Hilversum. The public broadcasters are all located in Hilversum, as are many production firms. The largest audiovisual producer of the Netherlands, Endemol, is located in Aalsmeer, to the south of Amsterdam. Amsterdam is also a very prominent location, in particular for firms active in the film industry and in the production of commercials.

3.1 Private Firms

The audiovisual sector in Rotterdam is very small and accounts for less than 1 per cent of total employment in the region. Statistics on the exact size of the sector are very scarce. The *Rotterdam Production Guide for Film, Television,*

Video & Multimedia mentions some 100 firms. The sector is dominated by relatively small companies: most firms employ fewer than five people. There is no single large audiovisual producer in the region. The audiovisual firms in the region predominantly work for commissioners in the business community or government agencies. The role of national and regional TV stations as clients for the Rotterdam-based firms is limited: although there are some producers who work for TV stations (national and/or local), their number is very small. The best known player in that respect is Blue Horse Productions, a large TV/video producer that moved from Hilversum to Rotterdam in 1997.

There are very few film production firms in Rotterdam. The region counts no large film producer: there are some very small films producers, employing only one or two people. Most of the Rotterdam film producers make 'cultural' films for relatively small audiences. Amsterdam is the most important city for the film industry, as a meeting place for the leading people in the Dutch film business. The same holds for producers/developers of TV and radio commercials. This category is also virtually absent in Rotterdam, with a few exceptions.

The scale of the audiovisual sector in the Rijnmond region is too small to support specialised services. In particular the number and scale of facility suppliers (cameras and other equipment) is very restricted. In many cases, Rotterdam-based firms have to 'import' their equipment from the Amsterdam or Gooi region when they work on a large production. An exception is the provision of video equipment, which is sufficiently represented. Other services such as casting and music/voice agencies are also virtually absent in Rotterdam.

3.2 The Most Important Customers

The demand side of the audiovisual cluster of Rotterdam can be subdivided in several categories. The most important are broadcasters, large firms/ organisations and government agencies.

Broadcasters Broadcasters are a very important part of the demand side of the audiovisual industry. The Rotterdam region counts no national broadcasting organisations: those are all located in Hilversum, Amsterdam and Aalsmeer. Even so, some of the Rotterdam-based production firms work on a frequent basis for national broadcasting organisations, both commercial and non-commercial. The Rotterdam region has a regional broadcasting organisation named TV Rijnmond, serving the Rijnmond region. The concept of the TV Rijnmond broadcasting organisation is to produce the lion's share of TV hours internally. To that end it has studios, facilities and many staff at its disposal. Only very few productions are outsourced to independent producers. TV Rijnmond is financed partly by the municipality of Rotterdam and the province of South Holland and partly by earnings from commercials.

A new development in the broadcasting landscape of Rotterdam is the recent establishment of Smartv, a small broadcasting organisation in the region. Smartv is largely aimed at (potential) students, both young people and adults, to inform them about education possibilities and opportunities in the Rijnmond region. A second aim was to provide education through the channel, but this target proved to be too ambitious. Smartv is located at the so-called open channel of the Rotterdam cable network. This means that the channel is used by a wide range of other broadcasting organisations. Smartv is financed by educational institutes in Rijnmond.

Business demand The potential demand on media services and products from local firms is enormous. As shown in section 1, the Rotterdam economy counts many large firms. These firms, with their large and increasing need for communication with their clients, suppliers and staff, have a pressing need audiovisual and related services. Most of the businesses outsource their requirements for audiovisual products. Some of the largest firms in the region, such as Shell and Nationale Nederlanden, have their own audiovisual departments. In most cases, the function of these departments is limited to supporting presentations, and to producing video films for promotion or instruction. They are equipped with professional staff and facilities, but also frequently make use of other firms, freelancers or facilities providers, in particular when large productions have to be made.

The production of TV commercials or other large campaigns is generally outsourced to professional and well-known advertising agencies outside Rotterdam.

Government agencies A third type of important client for audiovisual products is government agencies. In Rotterdam, the most important ones are the Rotterdam City Development Corporation, the Port Authority and the housing department. They use audiovisual productions for such purposes as internal and external communication and information supply.

3.3 Education

The educational institutes form an important element of the audiovisual cluster.

The Rotterdam region counts several education institutes that offer education in the audiovisual field. The most important are the graphical high school, the design academy and the Ichthus school for higher education.

Graphical High School The Graphical High School is an education institute for vocational training. It counts 2,000 full-time and 1,000 part-time students. It offers a specialised education programme on technical aspects of production, camera handling, sound and vision. Some 25 students are currently enrolled in this course. In September 1998, the Graphical High School started a new type of course, aimed at the audiovisual design. From then on, the school offered a total package of audiovisual education. The audiovisual courses are very popular with students. It is expected that the total number of students in the two audiovisual courses will rise to 200 in a few years. In the basement of the building of the Graphical High School, located in the centre of the city, studios are being built where students can practise. These studios will contain state-of-the-art digital equipment.

Design Academy The Design Academy is part of HR&O, an institute of higher education. The academy has two types of major subject: applied arts (80 per cent of the students) and autonomous arts (20 per cent). The number of applied students is relatively high compared to other design academies in the Netherlands. Within the applied education, the academy offers programmes in audiovisual design and multimedia design: in total, some 60 students follow this course. The focus in this course is on the design of traditional (non-digital) audiovisual productions, but also on computer animations and manipulation. An important element of the courses at the academy relates to business education: students learn how to start a new firm and how to run a business.

A specific problem for the Design Academy is the policy of the national government of the Netherlands drastically to cut the budgets for arts education. From 1998 to 2001 the arts academies in the Netherlands have to reduce their total budgets by 17.5 per cent. This has stimulated them to find other ways to finance their activities. It is partly for that reason that the Design Academy is actively looking for contacts with the business community and principals to win contracts and to set up programmes of contract education for professionals.

Ichthus The Ichthus Institute for Higher Education offers a complete four year curriculum dedicated to communication. This course counts a total of 1,500 students. Each year, some 400 students graduate. Within this regular communication course, students can choose a course named 'creative communications'. The development of creative and artistic capabilities forms the core of this programme. Particular emphasis is put on the visual aspects of communication.

In due time, a new programme will be offered to students in applied media. Students will learn how to compose the right media mix for a given communication purpose. To that end, they need to know about communication in general but also more practically about the different media that are available. Like the Design Academy, Ichthus offers special courses in starting-up and running a business. This course has been developed in cooperation with the Rotterdam Chamber of Commerce.

Others There are other educational institutes that offer education aimed at the audiovisual sector. The Erasmus University has a media course as part of the Faculty of Arts and History. It also offers a postgraduate course in journalism.

3.4 Government Policies

The audiovisual cluster receives much attention from policy makers in the Rotterdam region, as the audiovisual industry is one of the policy spearheads of the Rotterdam City Development Corporation (RCDC). The RCDC regards the sector as a potential motor for economic development and job creation in a region that is characterised by relatively high unemployment and a one-sided economic structure. There are different projects. The most important are the film fund and the redevelopment of the Schiecentrale, a former power station, as a location for audiovisual industry.

The Rotterdam film fund The RCDC operates a film fund with a total budget of DFl 4 million. The aim of this fund is to stimulate the audiovisual sector in the region, to enhance the image of Rotterdam as a film city. Film-makers or producers receive loans from the fund to finance a production under the condition that they spend 150 per cent of the total loan within the Rotterdam audiovisual sector. For non-Rotterdam-based film-makers who make use of the fund it has proved very difficult to fulfil the obligatory expenditure in the audiovisual sector in the Rijnmond region, mainly because the sector is so small.

Development of the Schiecentrale The Rotterdam municipality's second major stimulation measure is the redevelopment of the Schiecentrale, a former powerhouse located southwest of the city centre in an old port area. It is intended to become the focal point of the audiovisual sector in Rotterdam: the municipality would like to cluster audiovisual firms in this building. Two large studios have already been built, as well as smaller sites and office space. Since 1997, one of the studios has been used by Blue Horse Productions, a large video/film producer that moved from Hilversum to Rotterdam. The RCDC actively tries to attract other firms to the area as well. The costs of around DFl 15 million were carried by the municipality.

Other policies The audiovisual industry is regarded not only as an economic sector: it is also one of the focal points of municipal arts policy. The Rotterdamse Kunststichting (Rotterdam Arts Foundation) pursues a policy of screenplay development and production stimulation. Its annual budget is DFl 150,000. This budget is mainly reserved for up and coming Rotterdam film and video makers who 'consciously want to keep operating on the fringe' (Het Initiatief, 1998). The contributions are of an initialising character and are not intended to supplement other subsidies.

4 Dynamics in the Cluster: Interaction and Spin-offs

This section is dedicated to the interaction and dynamics in the audiovisual cluster in Rotterdam. We analyse in what way and to what extent the different types of actor cooperate amongst themselves and with other actors, both within and outside the region. In Figure 10.3, the different relationships are depicted. The numbers correspond to the subsection numbers in which the interaction is described.

4.1 Interfirm Relations within the Cluster

The Rotterdam audiovisual sector is characterised by very dense interpersonal networks. The sector is so small that 'everybody knows everybody'. To a large extent, this is due to the project-like way of working in the business. Firms active in audiovisual productions often work with freelance people for separate projects. Most of the firms have a database of preferred freelancers they hire frequently. In addition, firms often pass jobs to colleagues when they are unable at that moment to engage in a project. Surprisingly, many Rotterdam audiovisual firms prefer to work together with other firms/people from Rotterdam.

Though interpersonal networks are well developed and firms pass jobs to each other, long term and strategic cooperation between firms is rare. In an

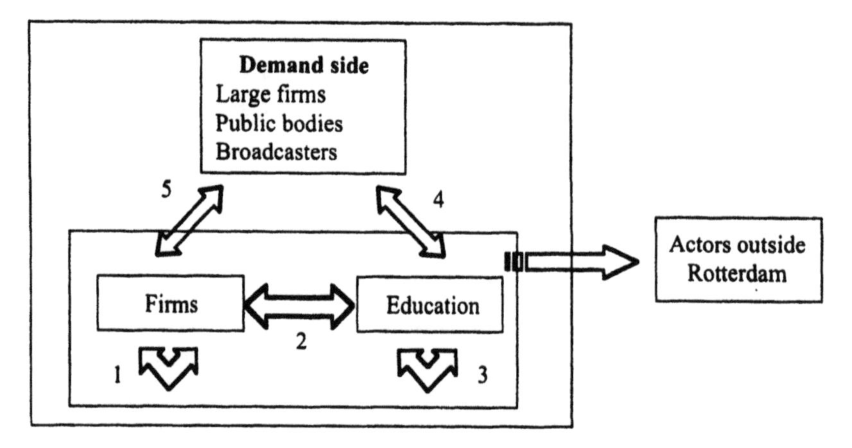

Figure 10.3 Interaction in the audiovisual cluster in Rotterdam

atomised sector with small firms, strategic cooperation might offer scope for development and growth of the sector in terms of quality and quantity. For example, firms might cooperate to carry out larger projects jointly, or approach and activate the local market.

The interaction between firms and people in the audiovisual cluster seems to have been increasing in the last few years. Since 1997, different types of firms and freelancers in the Rotterdam audiovisual sector have organised themselves in a branch organisation called 'Het Initiatief' (The Initiative), with 200 members. Originally, this organisation was founded because of discontent with the local government policy concerning the sector, but it also aims at enlarging the visibility of the audiovisual firms in the region. To that end it has published and distributed a booklet listing all members, with their competencies and activities.

External dependencies The audiovisual industry in Rotterdam is not a complete sector where all elements are present on a sufficient scale. There is a particular lack of equipment/facilities firms in the region. This is a severe impediment for many Rotterdam-based audiovisual production firms. In that respect, the firms are very dependent on facility firms in other parts of the country. For example, the most modern cameras have to be 'imported' from Amsterdam or Hilversum, which is relatively costly for the Rotterdam audiovisual producers. The general feeling is that the scale of the sector in Rotterdam is too small to support a sufficient facilities infrastructure. Furthermore, some firms blame the local broadcaster TV Rijnmond for having itself bought all facilities, thus significantly reducing the demand for facilities in the region. It is not only equipment that has to be brought in from other regions: in many cases, the same holds for human resources, such as directors and different types of crew. This kind of expertise is also very much concentrated in and around Amsterdam and Hilversum.

4.2 Interaction between Audiovisual Firms and Education

In the relation between audiovisual industry and the education sector, four aspects are important. First, the future employees of the firms are educated in education institute. In general, the audiovisual firms in Rotterdam are not entirely happy with the quality of the graduates from the education institutes. They often complain about a lack of practical skills and ability to handle modern equipment. Second, the firms are important for the on-the-job training of students; many audiovisual firms use students from one of the Rotterdam institutes, but also, in many cases, students from elsewhere are attracted. Third, in some cases the education institutes function as vocational training institutes for people who are already employed in the sector. This holds in particular for the Graphical High School. This type of business-education cooperation is, however, still relatively rare, even though, given the rapid (technological) developments in audiovisual production, it would be very beneficial for both the education institutes and the firms. The Design Academy, for example, provides part-time education courses, but the structure of the course is not adapted to the needs of people from the business. Fourth, education institutes are the breeding ground of potential new firms in the audiovisual industry and new media firms. Quite a lot of students start their own businesses, in particular from the Graphical High School and Ichthus. Most of the students start their business in Rotterdam, where cheap housing and space is available, and because of the highly appreciated dynamic climate of the city. This indigenous development is most important for the development and dynamics of the audiovisual/media cluster in the region, and deserves more attention as a vital focal point of integral audiovisual and media policies.

4.3 Interaction amongst Education Institutes

For many years, the relationship between the different educational institutes in the region was characterised by mutual distrust; as a result, cooperative structures were weakly developed. During the last few years, however, the climate has changed: cooperation between institutes is beginning to emerge, although still on a limited scale.

In the field of audiovisual (and related) education, there are different cooperation initiatives between the education initiatives active in the audiovisual cluster in Rotterdam. Firstly, the Arts Academy plans to set up a new course together with the Graphical High School in the field of graphics and (new) media. Secondly (not in direct relation to the audiovisual cluster), the Design Academy is working on a new course together with the informatics sector of the HR&O higher education institute. This project should result in a very interesting combination of design and computer technology.

The level of strategic cooperation amongst educational institutes could be much improved. Currently, the programmes of the different levels of audiovisual education are not adapted or fine tuned. A better coordination of media-related education on different levels in the region could raise the quality of education and make Rotterdam more attractive for students who are interested in media education. There are prospects in the streamlining of education programmes and permitting circulation between levels and institutes. Moreover, joint marketing efforts of the institutes, perhaps even in cooperation with firms active in the media, may put Rotterdam more firmly on the map as a media education centre. This, in turn, might further enhance the potential economic spin-offs from education for firms in the cluster.

4.4 Interaction between Education and Demand Side

The Design Academy has started a project with the title 'Kunst in Opdracht' ('arts on demand'). This project links students and firms in the region: students' artistic productions are bought by firms/organisations in the region. This project is innovative, because for many years there were virtually no relations between artists and the business community: arts were considered to be independent and autonomous and links with business were suspect.

4.5 Interaction between the Demand Side and Audiovisual Firms

The interaction between production firms and principals is generally characterised by trust and continuity. Often, principals come back to the firms they have worked with. Furthermore, frequent face-to-face contacts are important in the cooperation between the audiovisual producer and the principal, as the making of a presentation, video or film for information requires intensive contact amongst the producer and the principal. This does not imply that geographic proximity is essential.

Large firms in the region Many large Rotterdam firms outsource their audiovisual productions to firms in Hilversum or Amsterdam. The main reason is the lack of scale and visibility of the audiovisual sector in Rotterdam: large businesses are almost unaware of the presence of audiovisual producers in the Rotterdam region. In addition, large firms often have communication demands that far exceed the capacity and capability of the small firms in the Rotterdam region. For example, when Shell Rotterdam wants to organise an information campaign for its personnel, it needs a perfect presentation with the largest screen possible. The firms in Rotterdam cannot deliver that type of product.

Government agencies Local government is an important client of the Rotterdam audiovisual industry: it requires a great deal of promotion and information films, audiovisual presentations, both for internal and external use. Many different bodies within the municipality have their own departments that are responsible for the outsourcing of audiovisual projects. The most important are the Rotterdam City Development Corporation (RCDC), the housing department, and the Port of Rotterdam. There seems to be little coordination between the different bodies regarding the outsourcing of audiovisual productions.

Local broadcasters The most important local broadcaster, Rijnmond TV, plays a minor role as a client for the audiovisual sector in the Rotterdam region. As a result, the potential function of the local broadcaster as engine behind the development of the cluster fails to materialise. In the ideal situation, a local broadcasting station provides a regular and constant flow of orders into the local audiovisual firms and functions as breeding ground for young talent. The educational broadcaster Smartv is too small to play a major role as a client for the sector in Rotterdam. Apart from that, most of the programmes are made by the Smartv team itself.

5 Confrontation with the Framework

In this section, the audiovisual cluster is confronted with the framework of reference. This framework puts the cluster in the perspective of the urban context, and analyses the organising capacity with respect to the cluster.

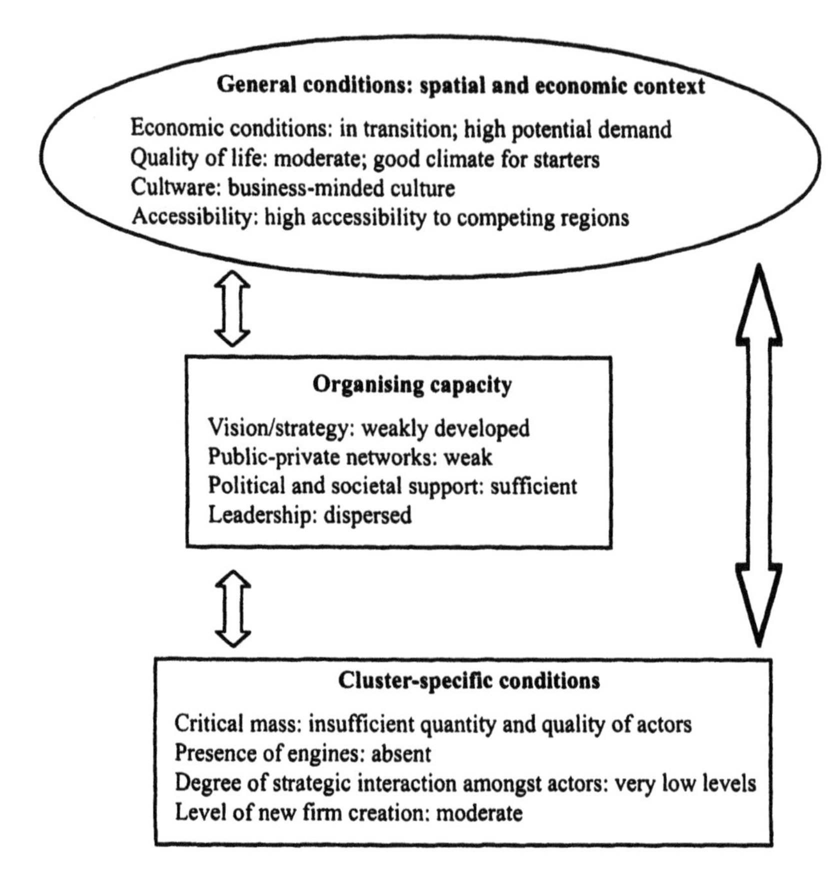

Figure 10.4 Framework of reference

5.1 Spatial and Economic Conditions

Although the Rijnmond region counts only a few real decision-making centres, there are very many firms and there is a lot of economic activity. This can be considered advantageous for audiovisual firms, because of the high (potential) demand for audiovisual products. Furthermore, this demand can be expected to grow because of the general tendency in the business world of growing importance of communication with clients and suppliers. The strong economic base of the Rotterdam urban region can act as an engine behind the development of the audiovisual cluster.

Regarding the 'cultware', traditionally, Rotterdam is regarded as a 'working city', even though its unemployment rates are higher than in the other large cities in the Netherlands. The no-nonsense mentality of Rotterdam can be found back in the audiovisual sector, in different respects. Rotterdam accommodates virtually no entertainment industry: by far the largest part of the Dutch TV and entertainment producers are located in the 'sparkling' city of Amsterdam, or nearby, in the traditional media area Het Gooi. With a few exceptions, the Rotterdam audiovisual sector is mainly business-oriented: business-to-business communication and informative video productions are the basic products of the sector in Rotterdam.

Regarding the spatial conditions, the most important aspect is the excellent external accessibility of the Rijnmond region. The region is very well connected to important centres, both by road and by public transport. This contributes to the ease with which so many large Rotterdam businesses can outsource their audiovisual productions to large specialised audiovisual firms in Amsterdam and Hilversum, very nearby. This also makes it difficult to attract firms from that region to Rotterdam.

Another asset is the quality of the living environment. On the one hand, people in the sector indicate that housing conditions could be better, in particular in the city of Rotterdam itself. On the other hand, the supply of cultural facilities and nightlife is judged positively. Very importantly, young starting entrepreneurs appreciate the fact that in Rotterdam cheap (office) space is available and that it is a dynamic city.

5.2 Cluster-specific Conditions

The strength of the cluster and the interaction between the different elements have been extensively discussed in sections 3 and 4. To summarise, the private firms in the audiovisual sector are small and dispersed. There is almost no film industry and facility providers and specialised services are virtually absent. Networks within the sector are very well developed and are becoming even more so with the foundation of Het Initiatief. However, real strategic networks, for instance with the aim of co-producing large productions or joint marketing efforts are as good as nonexistent. Education facilities in the field of audiovisual are relatively abundant in the region, considered the very small size of the sector. However, cooperation amongst and between institutes and the audiovisual sector could be improved.

5.3 Organising Capacity

In general, the organising capacity regarding the audiovisual cluster shows several shortcomings. In the first place, there has, for too long, been a lack of vision about the way the audiovisual sector should develop in relation to the possibilities in the context of the urban economy of the city, the characteristics of the audiovisual cluster and the trends in the communications industry. Consequently, there is no clear and balanced strategy on the development of the cluster. As a result, too much effort has been put into ad hoc policy measures, such as attracting existing firms from outside the region.

In the second place, the private sector has been insufficiently involved in the design of policies for a long time. Civil servants mainly developed plans and initiatives to stimulate the audiovisual cluster, for instance the development of the Schiecentrale and the establishment of the Filmfund, without broad consultation and involvement of the private sector. This has resulted private entrepreneurs developing a negative attitude towards the municipality. It is also felt that investments in the cluster could have been more effective if there had been a clear strategy and support from the private sector from the beginning.

Another important aspect is the lack of leadership towards development of the cluster. Illustrative in that respect is the dispersed attention within the municipality. On the political level, attention is divided between two aldermen: one for economic affairs and one for cultural affairs. This hampers the effectiveness and coherence of policy measures. On the level of public administration, too, there are coordination difficulties. Within the Rotterdam City Development Corporation, several people work on the cluster, but coordination too often falls short. A complicating aspect is that other departments of the municipality are involved in media policies as well. The most important is the arts/cultural department, which distributes the subsidies amongst the arts institutes.

It would be effective to develop a separate decision-making structure on which representatives from different departments have a seat, together with delegates of the private sector and education institutes. This new entity should take care of the development of an integral and broadly supported strategy and should be able to invest the funds available for the development of the cluster accordingly. It is important to keep a broad perspective, implying that focus should not be on the audiovisual sector alone but also on the new media in which audiovisual productions are incorporated to a growing extent.

6 Conclusion and Perspectives

The communications sector is becoming a vital industry, as the need for all sorts of communications with businesses increases. Rapid developments in information and communication technologies reinforce this development, and offer a broad range of new opportunities. Given the fact that audiovisual products increasingly become an integral part of 'new media' productions, the emergence of new media can be expected to offer a substantial growth potential for audiovisual productions. In that perspective, the audiovisual industry is not just a single, separate branch, but forms an important supporting and 'enabling' industry for the new urban economy as a whole.

The cluster perspective as presented in this study can be a helpful device to build on new strengths and develop the audiovisual cluster in a coherent way. Both the linking of demand and supply and the formation of strategic relationships between the various actors in the cluster – aimed at enhancing the visibility of the sector and the improvement of the quality – are essential. As a result, large clients in the region may eventually satisfy part of their media demand in the Rotterdam region instead of in Amsterdam or Hilversum, although the effect should not be overestimated, given the current small size of the cluster. The coupling of supply and demand is essential to guarantee a constant stream of orders for the audiovisual industry. Potential engines in the region to provide for this are the previously-mentioned large firms, but also the local broadcasting organisation. Better utilisation of this demand potential might increase the quality and the quantity of the audiovisual cluster and strengthen the urban economy. Note, in this context, the importance of more in-depth research to determine under what conditions these potentials can be exploited.

Very importantly, the educational institutes in the region offer much potential for the media sector in Rotterdam, as 'knowledge factories' and as creators of talent in the region. They create dynamics as a breeding ground and source of new firms in the fields of audiovisual-related activities. Furthermore, they provide trainees and staff for existing media firms and principals. The spin-offs from education can be strengthened further if education institutes work together in coordinating programmes, sharing facilities and combining marketing efforts. This might help to offer better opportunities for students to specialise and may put Rotterdam firmly on the map as the location for media education.

For the future development of the cluster, it is fundamental to keep (young) talent in the Rotterdam region. This requires the creation of specific, targeted conditions regarding starters support and accommodation.

References

Booz.Allen&Hamilton (1998), *Benchmarkstudie Electronische Diensten: 'Op weg naar de informatie-maatschappij'*, internationaal vergelijkend onderzoek t.b.v. de herijking van het actieprogramma Elektronische Snelwegen.

Castells, M. (1996), *The Rise of the Network Society*, Blackwells Publishers.

Chamber of Commerce (1995), *Industrie in de regio Rotterdam: Plan van actie voor de regionale industrie.*

COS(1996), *Statistisch Jaarboek 1996.*

Croteau, D. and W. Hoynes (1997), *Media/society: Industries, Images and Audiences*, Pine Forge Press, London.

European Union (1993), *White Paper: Growth, Competitiveness and Employment: The challenges and Ways Forward into the 21st century.*

Fidler, R. (1997), *Mediamorphosis: Understanding new media*, Pine Forge Press, London.

Filmfonds Rotterdam (1997), *Samenvatting van drie rondetafelgesprekken georganiseerd door het Filmfonds Rotterdam april-september 1997.*

Het Initiatief (1998), *The Rotterdam Production Guide for Film, TV, Video & Multimedia.*

Klink, H.A. van (1996), *Towards the borderless mainport Rotterdam*, dissertation, Tinbergen Institute.

KPMG (1996), *Rotterdam Consolidated Top 100*, KPMG.

McKinsey (1993), *Stimulating Audiovisual Production in the Netherlands*, Audiovisueel Platform.

Nederlands Economisch Instituut (1994), *De audio-visuele sector in Rotterdam: Economische betekenis, ontwikkelingsmogelijkheden en effecten Filmfonds*, NEI.

Ontwikkelings Bedrijf Rotterdam, *Trendbericht 1997: Sociaal economische rapportage*, OBR.

Rotterdam Media Agenda (1998), *Praktijk en beleid op het gebied van kunst en cultuur en digitale media in Rotterdam.*

Smits, B. (1998), *Rotterdam en de ontwikkeling van de AV-sector*, Audax Tros Multimedia.

Wolff, J.P. (1998), *Production is key in the film industry*, dissertation, Erasmus University.

Discussion Partners

Mr P. ten Arve, Rotterdam City Development Corporation.
Mr A. van Boven, Graphical High School, college van bestuur.
Mr G. van Collenburg, Smartv, Director.
Mr A. den Draak Den Draak Animation Studios, Director.
Mr S. Jonker, Rotterdam City Development Corporation.
Mr M.W. Knoester, Ichthus Institute of Higher Education, Course Manager Communication.
Mrs M. van Leewaarden, Rotterdam Films Production.
Mr J. van Nierop, Nationale Nederlanden, Audiovisual Productions.
Mr R.E. Ouwerkerk, Academy of Art & Design, Head of College/Executive Director.
Mr D. Rijneke, Rotterdam Films Production.
Mr P. Schuiten, Peter Schuiten Productions.
Mr B. Schuite, Shell Netherlands BV, Audiovisual Production.
Mr M. van Staveren, Men at Work RTV Productions.
Mr S. Warmenhoven, Het Kader Audiovisual Productions.

Chapter Eleven

The Health Cluster in Vienna

1 Introduction

The health cluster has a long tradition in Vienna: the city once was world-famous for its medical school and its top-level research. Currently, health care and the medical industry are at the centre of attention in Vienna. The Vienna-based hospitals, pharmaceutical and medical firms, medical and biological research institutes and other health related activities are increasingly regarded as powerful assets of the city, with a substantial economic impact and potential. At the same time, it is recognised this potential is not being fully realised.

It is against that background that this case study analyses the health cluster in Vienna. The cluster perspective is used to detect the presence (or absence!) of strategic relations among actors in the cluster and has proved to be a powerful analytical tool to determine cluster dynamics.

The study is organised as follows. Section 2 contains a brief introduction into the economy of the city of Vienna. Section 3 introduces the main actors and their activities in the cluster. Section 4 takes the analytical view, as it describes and analyses relations amongst and between actors and pays attention to new firm creation. In section 5, the cluster development is put in the context of the general urban economic development and organising capacity, in line with the general framework of the growth clusters study developed in Chapter Two. Section 6 concludes.

2 Profile and Economy of Vienna

2.1 Profile

Vienna, the capital of Austria, is located in the eastern part of Austria, and counts 1.6 million inhabitants, 20 per cent of the total Austrian population. In 1996, the capital accounted for approximately 29 per cent of the Austrian Gross Domestic Product, and employed 21 per cent of the total workforce of Austria (VBPF, 1997).

Politically, Vienna has a particular status in Austria, as it is not only a municipality, but also a federal province. The mayor is also head of the Provincial Chamber, and the city Senate is also the provincial government. The City Council has 100 members.

Vienna's role and position in Europe have changed fundamentally in the last decade. Firstly, the fall of the Iron Curtain had a great impact: from a relatively peripheral position in Western Europe during the Cold War, Vienna has now gained a very central location at the crossroads of Western Europe and the former Eastern bloc countries. Secondly, Austria became a member of the European family by joining the European Union in 1995.

2.2 Economic Structure

With a contribution of 28.6 per cent of the Austrian Gross Domestic Product, Vienna is by far the most important economic centre of Austria. The average unemployment rate is low at only 4.9 per cent, particularly by European standards (WWFF, 1997).

The economic structure of Vienna is dominated by service activity, representing 75 per cent of jobs. Many services in Vienna are related to the city's function as national capital, but private services such as banking, insurance and business services also take up a substantial part of the Vienna employment. The industrial sector is much smaller and generates some 24.6 per cent of employment. Most of the industry is knowledge-intensive. Low-tech industry has almost disappeared from Vienna (WIFO, 1996).

Vienna is an important location for corporate headquarters. Of the top 500 Austrian firms in terms of turnover, 203 have their headquarters in Vienna, controlling more than 50 per cent of the turnover of the Austrian top 500. In the European perspective, Austria's role as decision-making centre is also considerable (WIFO, 1996). Particularly interesting is the strong position of Vienna as a trading and service centre for the former Eastern bloc countries: Vienna has a relatively high density of multinational subsidiaries with commercial activities in Eastern Europe. In the future, this might reinforce the weight of Vienna as a European control centre.

2.3 Economic Development

The Vienna economy has undergone major changes in the last few decades. As in most other European cities, the service sector has grown. Table 11.1 shows the sharp decline in employment in industry and the growth of the service sector. Table 11.2 gives a more detailed account of developments within industry. On average, 33 per cent of the jobs in industry were lost between 1981 and 1993. Food, paper and metal and chemical industries, with losses exceeding 40 per cent, are extremely hard hit. The added value in most industry sectors increased considerably, indicating a fast growth of productivity of the Vienna industry.

Table 11.1 Industry in Vienna: employees and added value in 1993, % change 1981–93

	Employees		Added value (ATS)	
	1993	% change, 1981–93	1993 1981–93	% change,
Food industry	7,981	-43.7	6,749,656	+19.9
Drinks; tobacco	2,760	-1.1	4,578,670	+56.0
Paper	2,088	-44.7	1,242,415	-9.7
Chemical industry	7,176	-29.1	5,809,066	+40.7
Metal	2,093	-57.8	1,204,018	-20.2
Machine building	7,299	-14.0	4,932,345	+67.7
Electrotechnical installations	30,675	-24.0	22,370,864	+88.6
Transport equipment	9,704	-19.1	6,594,476	+92.7
Construction	2,494	+16.2	1,309,632	+112.8
Total industry	86,300	-33.7	65,339,703	+46.4

Source: WIFO, 1996.

Table 11.2 Economic structure of Vienna, 1973, 1986 and 1994

	Total number of jobs	Primary sector	Industry	Services
1973	752,877	0.5%	40.1%	59.4%
1986	731,871	0.4%	29.4%	70.2%
1994	773,068	0.4%	24.6%	75.0%

Source: WIFO, 1996.

While industrial employment declined, service employment grew by 133,000 persons in the period from 1973 to 1995. Some 75 per cent of Vienna jobs currently are in this sector. Note that the shift from secondary to tertiary activity is partly a statistical phenomenon: many industrial firms have outsourced service activities, causing a mere statistical shift.

3 The Health Cluster in Vienna

3.1 Introduction

Due to technological, demographic, social and economic developments, the health sector (or medical sector) is growing very fast. For an account of the most important drivers behind this growth, we refer to Chapter Seven, section 3. To what extent does the city of Vienna benefit from developments in the health industry? To answer this question, insight is needed into the constituents and dynamics of the health cluster in Vienna.

Table 11.3 Employees in the health cluster

Type of actor	Employees
Hospitals	32,295
Pharmaceutical Industry	4,200
Medical technology (broad definition)	900
University clinics/institutes at medical faculty	720
Other university institutes	500
Total	38,615

Sources: Clement et al., 1998: ÖBIG, 1998; WKV, 1998.

This section describes the main 'actors' of the Vienna health cluster. For our purposes, we will select the following actors:

- hospitals (academic and non-academic);
- firms: pharmaceutical industry (research and production) and medical technology firms;
- public knowledge infrastructure: higher education (universities: science and medicine) and university research institutes (with clinical research and biotechnology research);
- government agencies (the city of Vienna, national government).

In terms of employment, the health cluster constitutes a very substantial part of the Vienna economy. Using a fairly narrow definition – many categories are excluded, such as the pharmacists, dentists, doctors – the cluster provides 5 per cent of Viennese employment.

3.2 Firms

In the Vienna health cluster, several types of firms – with respect to the activity they are engaged in – can be discerned, though it is not always easy to make clear distinctions between firms. The most convenient and straightforward method is to identify four categories: the pharmaceutical industry; biotechnology firms; medical technology firms; and a 'the rest' category, including, for instance, medical informatics and robotics. However, large pharmaceutical firms often have a biotechnology department as well, for the research and development of new products. Therefore, we will treat biotechnology firms separately only if they are small and more or less independent from the large pharmaceutical firms.

Large pharmaceutical firms are probably the most important category. Although there is almost no pharmaceutical industry of Austrian origin, many foreign pharmaceutical multinationals have a subsidiary in Vienna. Some of them only serve as selling outlets, but others also carry on production and research. In Table 11.4, the principal Vienna-based pharmaceutical firms are depicted.

Table 11.4 Large pharmaceutical firms in Vienna

Name	Activities	Number of employees 1996	Turnover (ATS, millions)	Mother firm
BI Austria	Production, sales, R&D	525	800	Boehringer Ingelheim
Hoechst Austria	Production, sales	575 *	3,330 *	Hoechst, Germany
Kwizda	Production, sales	580	2,000	Private person
Novartis Pharma	Production, sales R&D	170	1,150	Novartis, Switzerland
Immuno***	Production, sales, R&D	2,000	6,936	Baxter, USA
Pasteur Mérieux Connaught	Production, sales	47	328	Pasteur, France
Pharmacia and Upjohn	Production, sales	98 **	527 *	P+U, Sweden

Sources: ÖBIG, 1998; * ÖBIG, 1994; **ÖBIG, 1995; *** Immuno.

Three firms are active in research and development in Vienna: Boehringer Ingelheim Austria, Novartis and Immuno-Baxter. They deserve extra attention, as they are not only selling outlets but fulfil a much more strategic role in the cluster. *Boehringer Ingelheim Austria* carries on top-level research in the field of molecular biology (in particular cancer) and cell biology (proteins). The firm is not active in basic research, but picks up where basic research ends, mainly in new drug discovery. BI's core competencies are in the field of oncology, molecular genetics and genomics. One hundred and sixty people work in the research department; 70 in production and distribution. The firm's aim is to be innovative, fast and cheap. To achieve this, BI has greatly improved its 'production process' of new drugs discovery. For instance, with the help of new methods, BI has managed to increase the number of biochemical tests from 50 to 50,000 weekly. Partly as a result, the development time for a new drug has decreased from 12 years in the 1970s to seven years at the time of writing.

BI works closely together with the Institute for Molecular Pathology, a fundamental research institute owned by Boehringer Ingelheim. Since 1985, this institute has been located at the Vienna BioCentre, together with university institutes with which it also cooperates (see section 5 for a more detailed account). Fundamental scientific results of the IMP are exclusively owned by Boehringer Ingelheim, and are elaborated further by BI's Vienna-based subsidiary.

The second large firm with strategic relevance for the cluster is *Novartis,* a merger of pharmaceutical giants Cyba Geigy and Sandoz. In Austria alone, this firm employs 2,400 people. One of the eight research centres of Novartis, specialising in dermatology research, is located in Vienna. The centre employs 250 people (among whom 220 are research staff; 20 per cent are foreigners). Interestingly, since the merger of Sandoz and Cyba Geigi, Novartis operates a fund worth SF 100 million, with the aim of financially supporting Novartis staff who want to start a new business.

The last large pharma-player in Vienna is *Immuno*, formerly 100 per cent Austrian but since 1996 part of the USA-based firm Baxter, a global player in the pharmaceutical industry. The research activities of Immuno Baxter are mainly directed to bloodproducts and vaccines. The R&D efforts of Immuno have risen from ATS 640 million in 1991 to ATS 762 million in 1995. Immuno has its head office, some research activity and three facilities for production in Vienna. Total employment in Austria amounts to 2,000. The majority of Immuno's research takes place 30 km from Vienna, where a research centre with 230 employees is located. For Immuno-Baxter, one of the advantages of locating in Austria is the closeness of expanding markets of Central and Eastern Europe.

Small biotechnological firms These form the second category of firms. They are relatively scarce in Vienna, as biotechnology in Vienna (and in Austria in general), in comparison to the USA, is strongly dominated by large pharmaceutical firms (OBIG, 1998). Examples of smaller biotech firms in Vienna are Intercel, a recently-founded firm at the Vienna BioCentre, and Nanosearch (a spin-off from the University of BodenKultur).

Medical technology firms This category of firms is quite diverse and ranges from protheses producers to developers of wheelchairs. Although clearly belonging to the health cluster, this subsector should be regarded separate from the biotechnology and pharmaceutical industry. The number of firms is restricted. Clement et al. (1998) count 18 firms in Vienna, employing 400 people. There are no large firms in this category present in Vienna. One of the larger firms is a subsidiary of Otto Bock, a world leader in prostheses. Otto Bock's headquarters are in Göttingen, Germany. Worldwide, the firm employs 2,800 people. The Vienna subsidiary is active in production and research: 130 people are employed, among whom 40 are researchers. In contrast to the small number of firms, much medical technological research is carried on in Vienna in this field: more than 200 research projects are running in the academic hospital and the Technical University.

Others There are several other firms that fulfil a relevant function in the cluster. An important and fast growing branch is medical informatics. In medical and biotech research, measurement and medical information processing with the help of computers is of growing importance. There are also rapid developments in tele-medicine. The best known firm active in this field is Siemens, with a specialised unit of 60 people for medical information systems in Vienna. Other relevant (but very small) branches are robotics and fine mechanics.

3.3 Public Knowledge Infrastructure

The public knowledge infrastructure, with its double role of education and research, fulfils a vital function in the health cluster as source of knowledge, innovation and educated staff.

A number of higher educational institutes are relevant for the health cluster. The most important are the University of Vienna (UoV), the Technical University (TU) and the Universität für Bodenkultur (BOKU). The University of Vienna is by far the largest university, and offers education in the fields of medicine, pharmacy, chemistry and biology (see Table 11.5). It hosts more

Table 11.5 University of Vienna: number of students and graduates in several fields, 1996/97

	Number of students	Number of graduates
Medicine	11,024	605
Pharmacy	1,352	119
Chemistry	869	64
Biology	3,013	139

Source: ÖSTAT, 1998.

than 50 per cent of Austrian students in these subjects. Additionally, it offers a course in veterinary medicine, with 2,174 students. The Vienna Medical School (part of the UoV) was very renowned in the period from 1800–1933; it was one of the world's leading research centres and attracted people from all over the world. During the Nazi period, and also in the first decades after the war, much of its fame was lost. However, there are signs that Vienna-based medical research and education can once again meet high international standards.

The Technical University (TU) is important for the health cluster as well, as it offers courses in interdisciplinary biomedical technology. Also, the TU offers a substantial course in technical chemistry (942 students in 1996/97 and 83 graduates). Finally, the Universität fur Bodenkultur (BOKU) is mainly an agricultural university. Even so, the university is very important as an education centre for the health cluster as it offers a curriculum in food and biotechnology, with 1,000 students enrolled.

University research institutes (with clinical research and biotechnology research) Many different research projects in the field of health and biotechnology run at the several universities of Vienna. A distinction can be made between clinical and non-clinical (or preclinical) research. The most outstanding non-clinical biomedical research location is the Vienna BioCentre, where five institutes of the University of Vienna are located (for molecular biology, molecular genetics, biochemics, microbiology/genetics and cell biology). In 1992, these institutes were moved to one location, close to the Institute for Molecular Biology, which is a private research institute of Boehringer-Ingelheim (see also section 3.1). The initiative for the relocation was taken by several national ministries, the city of Vienna (providing land and space), the University of Vienna and Boehringer Ingelheim. In total, some 400 researchers work at the campus. Another location where much preclinical biotechnology research is done is the University of BodenKultur, with several institutes (centres for ultrastructure research, genetics and applied microbiology).

The *clinical* research activities mainly take place at the university hospital of the Allgemeines Krankenhaus (AKH), a very large hospital with 2,178 beds and 7,032 employees. Several clinics are located there executing substantial research: the clinic for Immune dermatology, general dermatology, gynaecology, oncology, clinical pathology, cancer and haematology. Furthermore there is an institute for immunology.

University research is funded from several sources. Firstly, universities have direct basic funding from the national government. Secondly, they can obtain money from the national research funding funds: the 'Fonds zur Förderung der Wissenschaftlichen Forschung' (scientific research promotion fund) and the Österreichische Akademie der Wissenschaften (Austrian Academy of Science). Approximately one-quarter of all research funds of the Austrian national government is dedicated to research into medicine and the health system (Federal Ministry of Health and Consumer Protection 1996). Thirdly, universities can engage in contract research for the industry. During the last few years, the universities in Vienna (and in Austria as a whole) have been confronted with severe cuts in national grants for universities. This has entailed several problems: the salaries of young researchers are very low, making the university a less attractive place to work for young talent compared to private firms. Universities are forced to seek other sources, such as contract research or increased participation in (inter)national research programmes and consortia. Currently, new employees can be hired only if there is external finance.

3.4 Hospitals (Academic and Non-academic)

After the firms and the knowledge-infrastructure, hospitals are the third type of 'actor' in the health cluster. Vienna is by far the most important hospital centre of Austria. The city counts 27 public non-academic hospitals, which are managed and owned by the City of Vienna. These hospitals generate a total employment of over 30,000 people (see Table 11.6). One-third of the budget of the city of Vienna is destined to hospitals.

Vienna's largest hospital complex is the AKH (Allgemeines Krankenhaus), an academic hospital. This complex is not only the largest in terms of number of beds (2,178), but hosts also virtually all specialisations. It is financed by the national government and the medical faculty of the university of Vienna. The AKH is the main focus of clinical research of the University of Vienna.

The remaining hospitals encompass general hospitals, specialised hospitals – used for treating certain illnesses – and sanatoria, nursing homes and convalescence and maternity homes. Also, there are several private clinics.

Table 11.6 Figures on public hospitals and nursing homes of the Wiener Krankenanstaltenverbund, 1997

	Hospitals	Nursing homes	Total
Number of beds	9,461	6,164	15,625
Long-term patients	356,833	10,426	367,259
Care days	3,098,474	2,180,554	5,279,028
Ambulant patients	3,440,176	0	3,440,176
Employment	26,110	5,432	32,295

Source: WKV, 1998.

The hospitals are very dependent upon decisions of the federal government and the city of Vienna. On the whole, they have very limited freedom of action. Nor do they have any incentive to act entrepreneurial, as any benefits accrue to the city of Vienna, or even result in budget cuts. The AKH is a special case in this respect, being a state (and university) owned academic hospital. Plans are being developed to give the hospital more room to develop new activities.

3.5 Locations of the Cluster Actors

Spatially, the different actors of the health cluster are very much dispersed across the city of Vienna. This can be seen in Figure 11.1, where the most important actors are depicted. The distances between the actors are generally considerable. For instance, it takes at least 30 minutes by car to bridge the distance between the Vienna BioCentre and the AKH. One cause of dispersion is the lack of a specific spatial policy regarding the health cluster of the city of Vienna. The only clearly planned action was the establishment, after long (political) struggles of the Allgemeines Krankenhaus (AKH) as concentration of several separate hospitals.

A recent location development is the construction of the technology park on the Donau Island, close to the UN building. This park is to be integrated with housing functions and leisure. Regarding business, the park aims at high technology firms, as well as at some institutes of the technical university (mechanical engineering). The park should also attract biomedical technology firms and medical informatics.

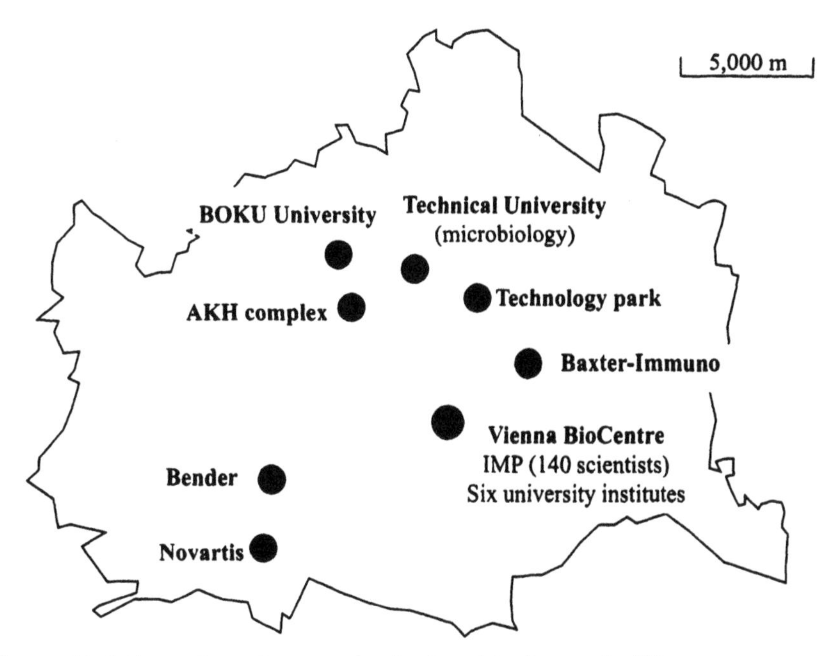

Figure 11.1 Location of actors in the health cluster in Vienna

3.6 Public and Semi-public Bodies (the City of Vienna, National Government, EU)

In the highly-regulated medical sector, public policies (on several levels) have a strong influence on its functioning. First, the city of Vienna can be mentioned. Some 10 years ago, the city of Vienna played an active and positive role in the strengthening of the medical cluster. It is generally recognised that the attraction of IMP (in 1985) and the formation of the Vienna BioCentre had much to do with the fact that the mayor was active in that respect. During the late 1980s and early 1990s, however, political action by the city regarding the health cluster (except for the establishment of the AKH) was very limited.

The city has no integral policy or strategic plan to stimulate the health cluster in Vienna. One difficulty is that responsibilities are dispersed among several councillors (at least three: health, economic policy, and planning). Nevertheless, the municipality is becoming increasingly interested in making more out of the health cluster economically. As a first step, it has commissioned several research projects to discover potential and steer policy decisions. An important role is played by the Wiener Wirtschaftsförderungsfonds (WFF), an economic stimulation fund owned by the city of Vienna with a budget of

ATS 400 million. Other actors are also involved, the Chamber of Commerce among them. One of the functions of this fund is to co-finance innovative projects or new firms in several fields, including, since 1997, biotechnology. Projects have to meet some conditions: they should be technically innovative and economically feasible, and should be based on a sound business plan. Furthermore, WFF never finances more than 50 per cent, with a maximum of ATS 20 million. At the time of writing, 18 projects in the field of biotechnology had been submitted. The projects are judged by a team of experts. WFF has 10 specialists in-house, but often expertise is hired. Since 1997, the WFF has taken initiative actively to stimulate entrepreneurship in biotechnology and medical technology, by visiting research institutes and informing scientists about the possibilities of new firm creation, financing and state aid.

On the national level, responsibilities regarding the health cluster are dispersed among several ministries. The Ministry of Science is responsible for the education of medical professionals and publicly financed research; the Ministry of Health should pursue general health policy (including law on genetic engineering). The Ministry of Economic Affairs promotes, among other things, industry policy aimed at the stimulation of biotechnology industry and firm creation. Owing to conflicting objectives and different views and interests, there is no coherent policy. Even so, there are a number of recent initiatives and policies relevant for the cluster. The first is the very recent plan to set up 'competence centres' in the next 12 years, with the aim of fostering the competitiveness of the Austrian economy by improving the links between public sector research and the industry. Competence centres are defined as collaborative research institutions aimed at high-quality pre-competitive and industrial basic R&D activities that fulfil the needs of the industrial sector and preserve high academic standards (Bundesministerium für Wissenschaft und Verkehr, 1997). They will receive up to 60 per cent public financing, but the rules for application are quite strict: for instance, contracts have to meet all kinds of specific condition. Because of these high levels of red tape, not all firms and universities in Vienna are happy with the new scheme.

The national government has a role in providing venture capital as well. It operates a venture capital fund called InnovationsAgentur Wien. This fund – exploited by a state-owned firm – finances innovations (or new firms) up to ATS 1 million and asks reimbursement only in case of success. Although the maximum amount is very small, finance from this fund greatly enhances the chance of getting grants from elsewhere, as the admission rules are very strict.

4 Dynamics in the Cluster: Interactions and Spin-offs

In the dynamic and fast-changing environment characteristic of the health cluster, rapid and frequent interaction and cooperation between the different actors in the cluster is of great value: a well-functioning cluster enhances flexibility, permits a better use of the available knowledge and boosts creativity in the region. Well-functioning strategic networks are likely to benefit all the parties, as they may generate a competitive advantage for the region as a whole. In that light, this section describes and analyses the strategic interactions between the actors in the cluster described in section 3. In Figure 11.2 the relationships that will be described are schematically represented. The subsection numbers correspond to the numbers in the figure. The last subsection deals with new firm creation, which is also a very important aspect of the dynamics in the cluster.

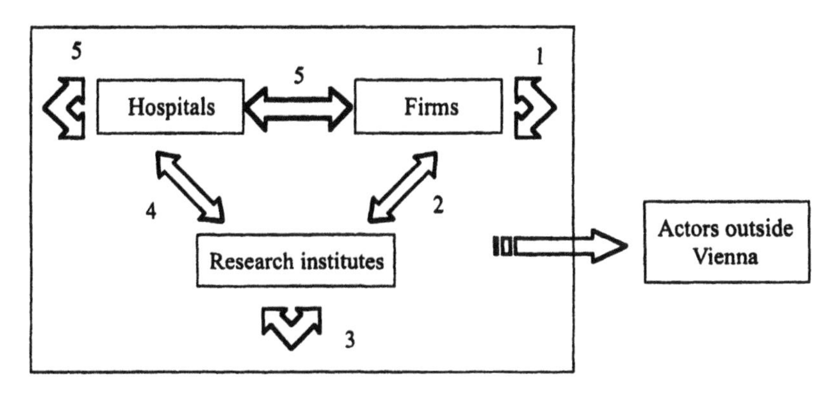

Figure 11.2 Relations in the cluster

4.1 Interfirm Cooperation

In this section, we will focus on the networks of the large pharmaceutical players, as they play such as strategic role in the cluster.

In the private sector in the health cluster in Vienna, strategic interfirm relationships do not prevail, and depend very much on individual firm's strategies. The large subsidiaries of pharmaceutical firms in the region (Novartis, Boehringer Ingelheim, Immuno-Baxter) are predominantly oriented to other divisions within the global conglomerates of which they are part. They often fulfil a specialised role. Even so, in the pharmaceutical industry some developments take place which might change the inward orientation of (some of) the Vienna-based firms. One is the trend towards 'lean organisations'.

In biotechnology and genetic engineering, where technological developments are so rapid, no single firm can master every field. Many firms increasingly choose to concentrate on core competencies, and engage in strategic relations with other, often small, firms that are complementary in terms of technology or markets. This enhances flexibility and gives the big players access to new technologies.

In that perspective, it is interesting that *Novartis*, one of the world's leading pharmaceutical firms, operates a special venture capital fund, to help Novartis employees who want to start a new business. To apply for funding, two basic conditions have to be met: first, there has to be an excellent business plan and second, the firm should not be active in Novartis's core business. The maximum contribution is 3 million Swiss francs. Worldwide, 30 spin-off companies from Novartis have already emerged. In Vienna, the first is about to be founded. Novartis is not the only case: other pharmaceutical firms have similar constructions.

The second large private player in the cluster, *Boehringer Ingelheim Austria*, attaches much value to strategic relations with small high-tech firms in the region, in particular in the field of biotechnology. But firms specialising in medical information technology are also very relevant, as Boehringer Ingelheim Austria needs to process and manage an enormous amount of information about test results. The same holds for firms specialising in robotics. Boehringer Ingelheim Austria indicates willingness to help start-ups in these fields in Vienna. Furthermore, there are examples of spin-off companies created by Boehringer Ingelheim Austria staff, although there is no structured policy to stimulate new firm creation.

Immuno-Baxter, finally, does not pursue such a 'satellite-strategy': the firm is more inclined to buy a majority share in any small firms active in its field. Furthermore, Immuno-Baxter does not have that many strategic relationships with other firms in the region, but maintains an ad hoc approach (for instance, influencing the government's biotechnology-related legislation).

4.2 Linkages between the Business Sector and the Research Institutes

Potentially, private firms and universities can reinforce each other by benefiting from each other's specialisations and capabilities. In Vienna, we discovered several types of strategic relationships between firms (active in biotechnology and medical technology) and public research institutes. In this section we will first consider industry-university relations and second look at spin-offs from universities.

The authors have the impression that strategic industry-university relations in research occur on an increasing scale in Vienna, although the level and number of partnerships is still limited compared to the situation in the USA. In that context, it is worth noting that contacts between researchers and the industry have always been present, but these relations have become more publicly accepted only recently and able to take place through more formal arrangements.

First, contract research activities seem to be taking root. One of the push factors for universities to engage in contract research is the continuing budget cuts by the national government, which put the finance of research in university institutes under severe pressure. Conversely, an important reason for pharmaceutical and biotechnology firms to seek strategic cooperation is the desire to benefit from expensive and risky fundamental research carried out there, since fundamental research results are often essential for applied research carried out in the private sector. Further, universities provide access to new knowledge and the brightest people.

Large pharmaceutical firms are particularly active in cooperating with public research institutes. One outstanding example of university-firm cooperation is the Vienna BioCentre, which houses a private research institute, IMP (Institute for Molecular Pathology) owned by Boehringer Ingelheim (BI), as well as six institutes of the University of Vienna (see also section 3). The decision to locate the IMP in Vienna was based on several factors. First, BI was by then engaged in a joint venture with Genentech, a USA-based biotech firm, which did not want the research department founded too close to BI, so that Germany was excluded as location. Second, a professor at Boehringer Ingelheim Austria in Vienna had good relations with Genentech, which made Vienna an attractive spot for the Americans. Third, the city of Vienna offered to contribute to the construction costs of the centre. Fourth, at that time a new chair was appointed at the University of Vienna, in the very field in which IMP was to become active, thus increasing the attractiveness of Vienna for the IMP.

Cooperation between IMP and the institutes takes several forms. There is a joint graduate programme of IMP and the University of Vienna; a joint seminar programme paid for by IMP, which attracts top scientists from all over the world; IMP contributes to the education programmes; and scientists from IMP lecture at the university. Furthermore, equipment is shared. Ideas are exchanged between people from IMP and the university, both formally, in all kinds of joint meetings, but, very importantly, also informally, for instance in the shared restaurant. It is generally recognised that the presence of IMP has raised the scientific level of the university, because of the presence of top researchers. The university has also taken advantage of the international status of the institute. Conversely, IMP benefits from the vicinity of the university institutes as sources of expertise and potential employees. There are plans to make the VBC a 'centre of competence', but the severe regulations and restrictions of the Ministry of Science are seen as a barrier.

Novartis participates in cooperation projects with university institutes as well, in particular in the Vienna International Research Co-operations Center (VIRCC), which concentrates on research in cancer and fighting infection. Novartis provides space, accommodation and administrative support at its Vienna location, while several university institutes (immune dermatology, immunology and physiology) carry the other costs. Novartis has the exclusive right to be the first to use new inventions (Clement et al., 1998).

Baxter-Immuno works with universities on a very limited scale. It operates a sponsor programme for the AKH, to finance the study of foreign students. In research, apart from the joint use of some facilities with the University of Vienna, Baxter-Immuno has no such intimate relationships with local university institutes but mainly with their own research department, located 30 km from Vienna. The main reason is the very specialised activity of the firms. Interestingly, since the merger of Immuno with the American giant Baxter, the regional networks of Immuno – the links with local universities and firms – amongst have lost importance: the firm is now more oriented to Baxter.

In the medical technology subsector, finally, other conditions hold. The number of firms is small (see also section 3.2), while public research – in particular at the medical faculty and the technical university – is fairly well developed. However, strategic interaction between firms and university institutes seems scarce. There are positive exceptions. For instance Otto Bock, active in production and development of prostheses, works with individual researchers from the TU on product development. An important problem for Otto Bock is the limited 'commercial awareness' of university researchers. For Otto Bock, cooperating in fundamental research – entailing a high risk – would be of great interest. However, the existing official support funds for business-university cooperation entail too high a degree of 'red tape' to be attractive.

4.3 Linkages among Research Institutes

Cooperation between researchers and research groups is essential for the development of knowledge, the exchange of ideas and a rapid diffusion of knowledge and innovations. Two types of cooperation can be discerned: intra-

and interdisciplinary. First, cooperation between teams with similar research programmes may yield critical mass and improve quality. Generally, in Vienna this type of exchange between research groups seems to be well developed, at least in the field of molecular biology, where three universities are active. In our interviews we found that, although research is dispersed among many institutes, there is no particular need for a further functional integration or spatial concentration of research in biotechnology, as (informal) networks of researchers are sufficiently developed. In the specific field of genetics, better cooperation between institutes could yield positive results, as this sector is dispersed and scattered. Furthermore, a joint graduate biotech programme between the various universities (BOKU, University of Vienna, Technical University) could enhance the quality of postgraduate education.

A second field in which cooperation is fruitful is where interdisciplinary approaches may yield interesting results, for instance medical and biotechnological research. Although there are many interdisciplinary initiatives – for example, the interdisciplinary cooperation project Molecular Medicine, in which several institutes participate – in some cases the situation in Vienna needs improvement. For instance, the research posts in the AKH are occupied by scientists with a medical background, in most cases practising doctors or specialists. In that respect, Clement et al. (1998) note that scientists with other backgrounds, for instance biochemists or gene specialists, are under-represented in the AKH research teams. As a result, interdisciplinary research activities in the AKH are exceptional, while in modern science this type of research is gaining importance.

Money shortages at the university institutes also hamper inter- (or trans-) disciplinary research projects, as these sometimes require high additional investments in new equipment. More support from the Ministry of Science to university financing would be helpful to overcome this.

4.4 Linkages between Public Knowledge Infrastructure and Hospitals

The hospitals in Vienna are important testing grounds for clinical research activities. This holds especially for the AKH, the large academic hospital, where many university institutes are located. However, the position of the AKH as a hospital for care and research sometimes causes problems. For some doctors, the time available for research is often restricted, given the scarcity of personnel in the hospital. Thus, the research potential at the AKH is not fully utilised. Another problem is that investment decisions are often delayed or complicated, because of the unclear division of responsibilities regarding the hospital between the city of Vienna, the university and the Ministry of Science.

Clinical research is not confined to the AKH. It also takes place at the non-academic hospitals in the city, though on a much smaller scale. In that respect, the Bolzmann Institutes play a role. They were erected to offer doctors in the regular hospital the opportunity to carry out research. The Bolzmann institute provides and funds the necessary facilities (labs, equipment). There are 60 Bolzmann research installations in Vienna.

4.5 Inter-hospital Cooperation

Cooperation between hospitals in the city of Vienna takes place on several levels and is mainly steered by the city of Vienna (owner of the majority of hospitals). For this task, the city operates the Wiener Krankenanstalten Verbund (WrKAV), a central organisation of the hospitals in Vienna. This organisation carries out administrative tasks for the hospitals, but also influences the degree of hospital specialisation. Furthermore, the WrKAV plays a role in innovation transfer, as they organise a yearly exhibition where new developments in health care are demonstrated. Further, there is cooperation on the level of information technology implementations. Finally, this organisation supports hospitals in introducing integral quality management.

4.6 Linkages between Hospitals and Firms

Hospitals and firms cooperate in several ways. First, the pharmaceutical industry and medical equipment producers need hospitals for the (clinical) testing of their products. In many cases, there are silent 'testing agreements' between individual doctors at the hospitals and pharmaceutical or medical technology firms. One problem is that individual doctors often have personal contacts with people from the established pharmaceutical industry (or medical equipment industry: Siemens, Philips Medical Systems), which makes them less willing to test products of new (and often small) firms. In many cases, hospitals levy a high charge for testing. Similarly, in many cases, creative ideas are sold to the large private firms which commercialise these ideas. Partly as a consequence, the number of medical experts in the hospital starting new businesses is small.

In preclinical testing there are also relations between industry and hospital, particularly the AKH. For instance, some of Novartis's products are tested by dermatology specialists from the AKH. Similarly, Immuno-Baxter makes frequent use of the AKH to test new (blood and plasma) products.

4.7 New Firm Creation

The creation of new firms is fundamental for the dynamics of the cluster, as new firms in the growth sectors of medical technology and biotechnology have many opportunities to develop and in the long run may reach high turnovers and generate new employment. Also, they may have an important strategic role for the already-present pharmaceutical giants – particularly Novartis and Boehringer Ingelheim Austria – which are likely greatly to benefit from the presence of innovative small-scale partners in their vicinity.

Unfortunately, however, the number of new firms created by university staff or students in the field of biotechnology and medical technology is low in Vienna. There are only a few examples of firms recently founded by researchers, such as Intercel (at the Vienna BioCentre). In Austria as a whole, only five firms were founded in 1997 in the field of biotechnology or molecular medicine. This is a low figure, not only compared with the USA but also in the European perspective (Ernst & Young, 1998).

This low number can be considered surprising, given the high quality of biotechnology research in Vienna even by international standards (Clement et al., 1998). The availability of venture capital has greatly increased in the last few years, with several public and private funds (see section 3), so lack of finance can no longer be blamed. Rather, the investment needed to start a biotech firms is not that high, as no expensive equipment is needed. Much can be achieved with ATS 5 million. More or less the same holds for medical technology. Several factors may explain the low rates of firms creation:

- firstly, scientific publication is considered more important than practical applicability of research results, as publication yields more scientific esteem than applications do. As a result, the number of patents taken out by scientists is fairly low;
- secondly, the entrepreneurial spirit of Austrian scientists is not very high. This can partly be ascribed to the prevalent negative attitude towards risk taking (characteristic of Austria and Europe as a whole, compared to the USA). This is reflected in the attitude towards business failure, which is considered normal in the USA but disastrous in Europe. The limited flexibility of the labour market also plays a role: in many cases, scientists's and doctors's contracts more or less imply lifetime employment and therefore remove any incentive for starting businesses;
- thirdly, there seems to be no adequate support structure for start-ups. Although there are several support organisations (among which is the

Wiener WirstschaftsFörderungsfonds, WFF), what seems to be lacking is a dedicated structure for bio- or gene-technological start-ups, that need very specific and professional legal, administrative and business support. The same holds for medical technology start-ups;

- fourthly, political factors seem to play a role. National and local government have considered new technology, in particular bio- and gene-technology, as a threat instead of an oppportunity for too long. This has certainly not contributed to a favourable climate for start-ups in this field;
- fifthly, specifically for medical technology and equipment, regulations in Europe – a so-called CE label is needed – are so strict that for small firms it is almost impossible to bring new products to the market. The risks are too high and time to market too long. At the same time, the position of established large suppliers of medical equipment is strong;
- sixthly, people employed by the hospitals cannot hold any patent. As they are civil servants, patents accrue to the city of Vienna or federal government. This greatly reduces the incentive to become commercially active.

5 Confrontation with the Framework

After the description of the main cluster elements and their interaction, this section puts the health cluster of Vienna in the analytical framework, placing it in the perspective of the characteristics and development of the city as a whole. The organising capacity regarding the cluster is also analysed. Figure 11.3 shows the framework.

5.1 Spatial and Economic Context of the Cluster

The health cluster functions in the context of a well-developed and diversified economy, with strong positions in virtually all sectors. For the future, the prospects of the city are good. In particular, the opening-up of Central and Eastern Europe may further reinforce the economic strength of the city, which is increasingly becoming a trading point between east and west. The location of Vienna at the new centre of Europe makes a clear strategic positioning necessary. On the one hand, the city must compete with the neighbouring former Eastern bloc cities, with much lower wages and a relatively skilled population. On the other, there is competition with Western European cities for high-grade activities such as corporate headquarters and R&D activities. The nearby city of Munich is a particular competitor. In the field of bio-

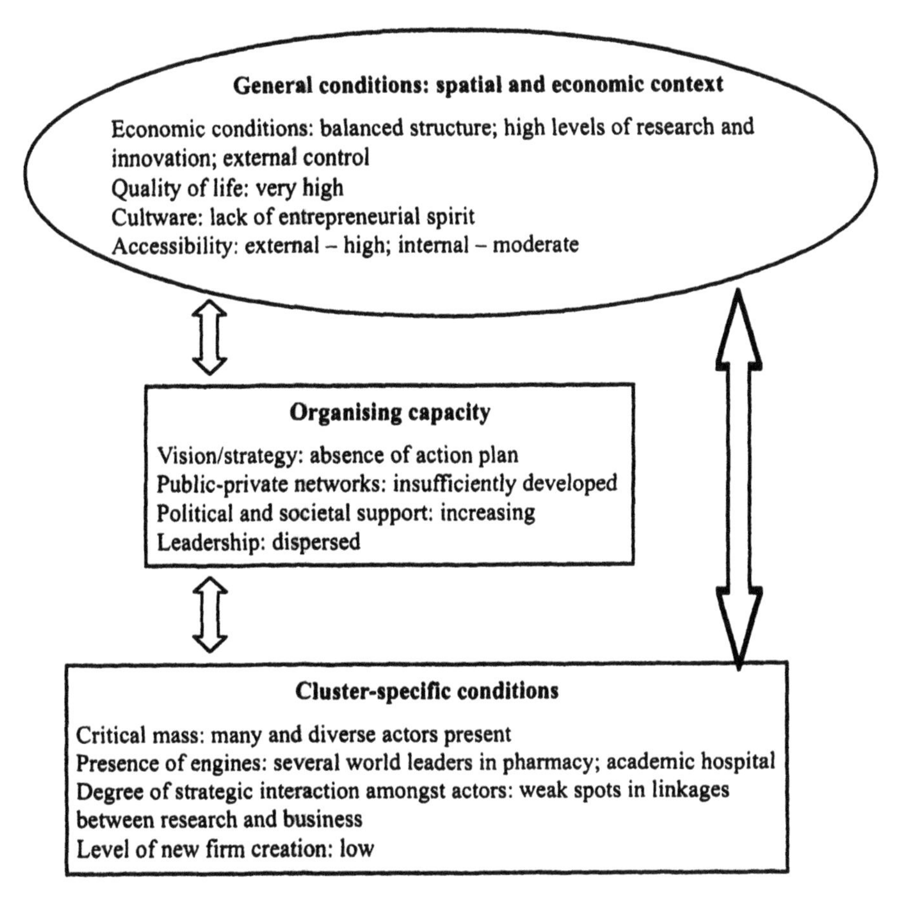

Figure 11.3 Framework of reference

technology, Munich is a powerful magnet for (new) firms and talent, as it has a very strong policy to stimulate these activities.

The high quality of the living environment in Vienna is an important asset for the city in general and a powerful one in the further development of the health cluster. The excellent cultural facilities – an enormous cultural heritage, a world-famous opera, museums – and the short distance to the countryside make Vienna a very attractive city for the researchers and other highly skilled people needed in the health cluster. Vienna is able to attract 'foreign brains', which is particularly important in the highly internationalised world of research and development. For foreigners, Vienna's cultural and environmental amenities seem to compensate the high Austrian tax rates. The attraction of foreign talent is further stimulated by the international orientation of Vienna in general

– there are many foreign firms and international organisations such as OPEC and UNESCO have their presence in the city – creating critical mass for foreign schools (an important location factor for the attraction of people from abroad).

With respect to the 'cultware' – the attitude towards innovation, willingness to take risks and do new things – there seems to be a general lack of entrepreneurial spirit, particularly in the health cluster. Several reasons can explain this: complicated regulations, a lack of incentives, the strong position of existing players – particularly true of medical technology, wide cultural differences (and mutual prejudices) between firms and universities, etc.

5.2 Cluster-specific Conditions

After this context description, let us summarise the cluster's strength. In section 3 and 4, two aspects of the cluster have been discussed: the actors in the cluster and their interaction. Summarising, we may say that the cluster has sufficient critical mass, with many hospitals, large and small firms and a huge university potential. Furthermore, the cluster has some 'engines', the most important being large firms: although owned by foreign companies, they are rooted in Vienna, and linked to some degree with the knowledge infrastructure and smaller firms. From section 4, the degree of interaction in the cluster needs improvement in some respects. There is no spatial concentration of the cluster: the actors are scattered throughout the urban region. Although long distances between actors and road congestion both hamper frequent interaction, this is not regarded as a serious problem, because the researchers know one another well. One the other hand, the success of the Vienna BioCentre points in a different direction: here, the physical proximity of university and the IMP has turned out to generate clear advantages.

5.3 Organising Capacity

The organising capacity regarding the cluster can be split into four elements: the presence of a vision and strategy regarding the cluster; the prevalence of public-private cooperation; the level of political and societal support; and the degree and quality of leadership.

Vision/strategy As stated already in 3.4, the city of Vienna has no strategic plan for the development of the cluster as a whole, although potentially the municipality has much power to influence the cluster. Policies are fragmented among different departments. For instance, the city-owned hospital organisation takes care of the hospitals, the WFF (Wiener Wirtschaftsförderungsfond) is occupied with private business development, location policies are directed from yet another department, etc. The lack of vision on the coherence and relations between the cluster elements hampers an integral, targeted action. The difference with Munich is striking: there, an integral and broadly-supported action plan was drawn up to give the biotechnology life sciences sector a strong impetus. Things seem to have change for the better, however, with the appointment of a cluster manager to develop the biotechnology and pharmaceutical clusters.

Public-private cooperation is indispensable to policy making and implementation. In the first place, as private actors, the large pharmaceutical firms in particular play such an important role as engines in the cluster that they should be involved in policy making processes. Furthermore, the knowledge incorporated in these firms about developments in the industry and specific needs is indispensable to effective and efficient policies. A second issue is the relation between the (semi-) public hospitals and business activity. The current unstructured relations (individual doctors deal with firms) may be profitable on the short term, but in the long run lead to reduced new firm creation and underexploitation of the research potential in the hospitals.

Political support for cluster development is slowly increasing, after a long period of 'laissez-faire'. Slowly but surely, politicians and civil servants are coming to recognise the enormous potential of the health cluster as economic engine for the city. This opens perspectives for a more structured approach and more targeted investments in the future. *Societal support* for the stimulation of the health cluster, particularly in the field of life sciences, is extremely important and deserves constant attention. Research in the field of genetic manipulation is regarded as threatening by a substantial part of the general public (although resistance seems to be slowly fading). For policy makers (but also for the industry) this implies that on-going communication and debate with citizens is essential to maintain support and to justify policy measures aimed at stimulation of these activities.

Leadership Who should take the lead in the cluster development? Large private firms take the lead in organising their own networks, and they do so quite well. At the same time, there seems to be a role for the municipality to take the lead in optimising 'missing links' between actors as indicated in section 4.

6 Conclusion and Perspectives

The increasing attention for and expenditures on health care and rapid technological developments offer great scope for the development of the Vienna health cluster as an economic sector. The existing cluster is already strong and contains large international firms, smaller firms, high quality public research and care. Weak points are the relatively weak position in medical technology, a high dependency on foreign firms, a very low level of new firm creation in promising fields and a lack of vision on the part of the municipality on the development of the cluster as a whole. Concerted action is needed, both to tie large firms to the city in a time of mergers and acquisitions and to withstand the pressure of other cities – notably Munich – as magnets for new development.

The cluster could become stronger and more dynamic if the municipality regarded the different elements as an interdependent whole and acted accordingly: the different elements can reinforce each other only if synergies are appropriately recognised. The direct influence of the municipality on the functioning of hospitals could be used to enhance the cluster dynamics, by providing hospitals with entrepreneurial incentives to enhance their innovative capacity.

The problematically low level of new firms creation and research commercialisation could be raised if a coherent and well-targeted starters policy were launched, with the ingredients to stimulate entrepreneurship at universities, offering specialised and professional services to new firms so that they can concentrate on what they are good at, and on creating the right environment. The involvement of private capital and know-how – particularly from the big firms – is one of the keys to success.

Past location policies regarding the cluster have left the city with the legacy of a very dispersed cluster from a spatial point of view, hampering interaction to some extent. New locational developments should be better planned, with the importance of face-to-face interaction between actors in mind. In the concrete case of the location of a possible starters facility, this means that close vicinity to research institutes is important.

The health cluster in Vienna has many ingredients to develop into a growth sector for the regional economy. Much needs to be done, however, to fulfil the ambition of the municipality to make more out of it economically by seizing opportunities and withstanding threats. Leadership and a vision on the cluster should ideally lead to a broadly supported integral strategy for the cluster a whole.

References

Arthur Andersen (1996), *Étude sur le potentiel d'attraction de la région Lyonnaise en matière pharmaceutique.*
Bundesministerium für Wissenschaft und Verkehr (1997), *Kplus: Forschungskompetenz plus Wirtschaftskompetenz.*
Clement, W., W. Kolb and R. Neuberger (1998), *Biotechnologie- Pharma- Medizintechnologie-Cluster Wien*, Industriewissenschaftliches Institut.
ERAI Entreprise Rhone-Alpes International (1998), *Réussir en Rhone-Alpes: les industries du genie biologique et medical.*
Ernst & Young (1998), *European Life Sciences.*
Federal Ministry of Health and Consumer Protection (1996), *Public Health in Austria.*
Financial Times (1998), 'Merger: Hoechst and Rhône confirm talks', 25 November.
NRC Handelsblad (1998), 'Chemiereuzen onderhandelen over alliantie', 25 November.
ÖBIG, Österreichisches Bundesinstitut für Gesundheitswesen (1998), *Biotechnologie: Pharmazeutische Industrie und Forschung in Österreich.*
ÖSTAT (1998), *Österreichische Hochschulstatistik 1996/1997.*
Tableau de l'Économie Francaise 1998–99.
VBPF Vienna Business Promotion Fund (1997), *Vienna Business Profile.*
WKV Wiener Krankenanstalten Verbund (1998), *Leistungsbericht 1997.*
WIFO (1996), *Wirtschaftsstandort Wien: Positionierung im Europäischen Städtenetz.*
World Health Organisation (1996), *European Health Care reforms, analyses of current strategies*, WHO, regional office for Europe, Copenhagen.
WWFF Vienna Business Promotion Fund (1997), *Vienna Business Profile, Information and service Handbook.*

Discussion Partners

Mr K. Anderle, Baxter-Immuno, Managing Director Plasma-Derivatives and Vaccines.
Mr W. Clement, Institut für Volkswirtschaftstheorie und -Politik.
Mr H. Dietl, Research and Development, Otto Bock Austria GMBH, Manager.
Mr K. Hosemann, Information Centre for the Vienna Economy, Wiener Wirtsschafts Förderungsfonds, Manager.
Mr Hutschenreiter, WIFO.
Mr F. Klopf, Economic Department, Wienerkrankenanstalten Verbund, Manager.
Mr W. Kolb, Industriewissenschaftliches Institut, Department Manager of Health Economics.
Mr H. Loibner, Novartis.
Mr R. Neuberger, Department of Education Economics, Industriewissenschaftliches Institut, Researcher.
Mrs I. Rosian, Österreichisches Bundesinstitut für Gesundheitswesen.
Mr E. Schillinger, Novartis Austria, Managing Director.
Mr U. Sleytr, Universität für Bodenkultur.
Mr P. Swetly, Boehringer Ingelheim Austria Wien, Managing Director R&D.
Mr N. Zacherl, Research Institute of Molecular Pathology (IMP), Administrative Director.

Chapter Twelve

Synthesis

1 Introduction

The objective of this investigation is to increase the insight into the dynamics of growth clusters in European cities. What factors affect the development of (potential) growth clusters? What can cities learn from one another's experiences? The starting point of this challenging international comparative study is the frame of reference that we have developed in the second chapter. This frame of reference identifies several factors that help to explain, analyse and compare (growth) clusters in the nine participating Eurocities, and to draw up policy implications for each of the cities as well. The previous nine chapters were dedicated to the case studies of growth clusters in the participating cities. At first sight, comparison seems difficult: firstly, the cases are dispersed across several countries, entailing country-specific aspects; secondly, the clusters differ in type, and thirdly, they differ in their 'development stage'. We analysed two mature health clusters (Lyons and Vienna), one very small media cluster (Rotterdam), a developing media cluster (Leipzig) and a very mature one (Munich), a large tourist cluster (Amsterdam), a specialised cultural cluster (Manchester), and two mature technologically-oriented clusters (telecommunications in Helsinki and mechatronics in Eindhoven).

Despite these wide differences, this chapter has the ambitious aim to bring the clusters together. Efforts are made to match the theoretical reflections of Chapter Two with the empirical evidence from the nine case studies and to draw some general conclusions and lessons about new growth opportunities and implications for urban economic policy. To put our frame of reference to the test and to make comparison possible, the structure of this chapter follows that of the framework of reference. Thus, section 2 synthesises the role of general economic, spatial and cultural conditions in the development of growth clusters. In section 3, the focus is on cluster-specific conditions. Section 4 draws conclusions on the role of organising capacity. The chapter ends with some final remarks.

2 General Conditions for Cluster Development: the Economic, Spatial and Cultural Context

The case studies show that the functioning, dynamics and opportunities of cluster development are largely dependent on the general economic and spatial conditions that prevail in the city under consideration. Cultural variables also seem to matter. In this section we will elaborate on each of these subjects.

2.1 General Economic Conditions

For most clusters, the general economic conditions in the city and region set the margins for growth as determinants of the demand for products that are produced by the cluster. This holds particularly for the media clusters, where substantial demand for media and communication products and services is generated by local firms. New media (and software) development in Munich flourishes, driven by the huge demand from powerful economic actors, whereas in Leipzig, the weak economic basis of the city implies a lack of demand for new media and software products. In the 'demand' for cluster products, the presence of headquarters of international firms, in their role of huge and critical demanders, proved to be important. In Munich, the presence of many headquarters (BMW, Siemens, Hypo-bank) is an important stimulus behind the development of media firms active in business-to-business communication. In some cases, for example the media cluster in Rotterdam, the demand potential of the region is felt not to be fully utilised. A policy implication is that stimulation of cluster development need not remain restricted to the cluster actors themselves: the activating of latent demand potential might in some cases be more effective. The economic sector structure of the urban region proved also relevant for the chances of cluster development. For instance, for the health cluster of Lyons, the large-scale presence of basic chemical industry is an important supplier of goods (and knowledge) for the health cluster. Third, the knowledge base existing in the city (the knowledge intensity of firms, the education level of the population) proved to be important, as all clusters under consideration proved to be highly dependent on the quality of the workforce.

2.2 Accessibility

In our analytical framework, we hypothesised internal and external accessibility as relevant factors in cluster development. From the case studies, we found that good internal accessibility – the ease with which actors in the urban region can get through to one another – enhances strategic cooperation in the cluster, as it brings cooperating actors nearer to one another and thus increases the chance of fruitful (new) combinations. However, it appeared that in many cases, the friction of physical distance is much less important than psychological barriers. Even the location of actors in the same building does not imply an incentive to cooperate. Personal contact seems to be a much more important determinant of cooperation than distance. Moreover, we found that proximity is positively related to the propensity to cooperate when the actors have 'grown up together' in the same building or location. An illustration of this is the in situ cooperation in the Vienna BioCentre, where the pharmaceutical firm Boehringer Ingelheim works closely together with institutes of the University of Vienna concerning fundamental and applied research. Another example can be found in Finland in the city of Olou, where very close ties between Nokia, smaller firms and the University of Olou have developed since they were located on the same campus. The ease with which other cities, national and international, can be reached by all kinds of modes – the external accessibility – is also relevant for the growth possibilities of clusters. The impact of the external accessibility depends on the type of city and the type of cluster. Clearly, good (inter)national connections make it easier for actors in the cluster to 'export' their products. It also increases the exposure of the cluster actors to international competition, which tends to make the cluster stronger. Owing to internationalisation of R&D and technological developments, international connections are indispensable to clusters in which technology and R&D are important (the health clusters in Lyons and Vienna, mechatronics in Eindhoven, telecom in Helsinki), to attract international staff, and to provide access to international partners. However, it is not just the technology-oriented clusters that put high demands on external accessibility. For the tourist cluster of Amsterdam, the strong position of Schiphol Airport is vital for its success in business tourism. Manchester airport could be instrumental to the international aspirations of the city's cultural enterprise. Good connections may have a negative impact on cluster development when strong competing cities are nearby. For Rotterdam, for instance, the proximity of 'media capital' Amsterdam makes it difficult to build up a media cluster of its own. The same holds, to some extent, for Leipzig, which competes with nearby Berlin in the attraction of several media activities. Another illustration is the cultural cluster in Manchester, where the attractiveness of London for creative talent is something to be reckoned with. Thus, cluster development in cities with strong 'magnets' in their vicinity will have to develop a clear specialisation based on local strengths instead of trying to emulate their neighbour. Urban specialisation becomes all the more relevant with the arrival of new fast transport means such as the high-speed rail network.

2.3 Quality of Life

The attractiveness of a city (in terms of housing, cultural and leisure facilities), proves a fundamental factor in cluster development, as a means to attract and retain highly skilled people to the region. In that respect it is interesting to compare the cities of Munich and Leipzig. Firms in the 'booming' media cluster of Munich manage to attract excellent staff from other German cities (and even from abroad) because of the superior quality of life that the city offers. By contrast, for Leipzig, with a much less favourable living climate, it proves very difficult to keep skilled people in the region, let alone to attract them from elsewhere. The unique quality of life and cultural amenities that many European cities can offer is also a weapon in the global competition for high-level staff. In Vienna, for instance, we found that for some high-level international researchers, the excellent quality of life in Vienna compensates for high income tax rates compared to other countries (notably the USA). Thus it can be said that Europe's heritage cities particularly are pearls of great economic value in the global competition for talent. Preservation and further amelioration of the quality of life is a long-term investment, with high pay-offs in the long run. The specific demands on the quality of the living environment differ by cluster. In the very technologically-oriented clusters – Eindhoven, Helsinki and, to a lesser extent, Lyons and Vienna – the quality of housing and the nearness of the countryside are considered to be important, while in the media-clusters (Rotterdam, Leipzig, Munich), as well as in the tourist (Amsterdam) and cultural (Manchester) clusters, the cultural climate and the metropolitan ambience appear to be somewhat more important.

2.4 Cultural Variables

Attitudes towards innovation, entrepreneurship and cooperation prove to be important factors in cluster development. The 'innovative culture' of a region, reflected in the willingness of people and firms to adopt new products and technologies and the willingness to experiment, again set the margins for dynamics and growth of cities. Firms in a cluster can benefit from an overall positive attitude towards innovation, as this brings about a 'home market' for new cluster products and an ideal testing ground. A strong and critical local demand for the cluster output gives cluster actors an incentive to remain cutting edge and provides opportunities for export. For instance, in the media cluster in Munich digital broadcasting techniques are tested in the very receptive local market. In Helsinki, experiments are run allowing ordering and paying for a can of Coca-Cola by mobile telephone at the city's airport. In Manchester the openness to cultural innovation is the basis for the cultural cluster development as such. On the other hand, the public concern in Austria concerning biotechnological research does not support the development of Vienna's health cluster. The attitude towards entrepreneurship is yet another relevant non-tangible cultural factor. Entrepreneurial people are indispensable to any cluster, as they bring about change. They are needed to discover new things, to make new combinations, to start new firms, and so on. We found very different attitudes in the several clusters. In Lyons, and to a lesser extent Vienna, entrepreneurialism was held in very low esteem by the universities, an attitude that hampers linkages between universities and business in the cluster. Partly as a consequence, the level of new firm creation is very low as well. At the other end of the spectrum are Eindhoven, Munich and Helsinki, where entrepreneurialism is more appreciated: students and teachers are much more inclined to link up with business and correspondingly higher numbers of start-ups and spin-offs from universities can be observed. The city of Leipzig is a special case, with a very low entrepreneurial spirit due to the legacy of communism. The municipality has even defined entrepreneurship as the leading principle of its economic policy and seeks to stimulate entrepreneurial activities. Although the attitude towards entrepreneurship is partly a cultural phenomenon, financial and legal incentives can do much to enhance it. In Vienna and Lyons, we found that entrepreneurial behaviour is rare because people have long-term, fixed contracts and virtually no incentive to do anything new. A decrease in direct financing may have the positive side effect of giving universities an incentive to execute contract research and seek contact with business.

Willingness to cooperate is a final key cultural factor of relevance, in a 'network economy' where access to resources of other organisations is vital. In the mechatronics cluster in Eindhoven, interfirm cooperation has been much facilitated by the high density of informal networks (sports clubs, unions, study-clubs, etc.). That 'informal basis' generates the necessary mutual trust that is indispensable for cooperation in innovative and risky activities. Ideally, cooperation emerges spontaneously, but policy makers could do much to create an environment that stimulates informal interaction. A good example can be found in Munich, were the municipality invested in the Literaturhaus (a meeting place for the publishing scene).

3 Cluster-specific Conditions

In our empirical analysis, with the help of our framework we studied several cluster-specific aspects: the importance of critical mass, the role of large companies as engines behind cluster development, the level of strategic interaction amongst cluster actors and the levels of new firm creation. Additionally, we found that the role of history and tradition can hardly be underestimated.

3.1 History and Tradition

Tradition and history matter in cluster development. Many cities included in the investigation have a tradition in the cluster we studied: for instance, Vienna (health) has always had a world-famous medical school; Lyons long served as the health centre for the whole of Southern France; Munich's function as an important media city (particularly publishing) dates back for centuries. The Amsterdam canals have been a tourist attraction since the early days of urban tourism and Manchester has gained a reputation in popular (youth) culture since the early 1960s. Tradition and history are the 'substratum' of many of the clusters in the investigation. From the cases, the clusters with a long tradition appear very well developed and complete. Tradition gives a lead, because often, history has created a valuable and well-established 'cluster infrastructure' that took years to build: a knowledge base, education institutes, research units, branch unions and so on. The social-cultural infrastructure in a cluster is of great value, as it determines the levels of mutual trust and willingness to cooperate, but it takes much time for such an infrastructure to come into being. On the other hand, an absence of history and tradition in a certain field of activities makes it very difficult to develop a cluster. This has become clear in the case of Rotterdam, where it proves to be extremely difficult to develop a media cluster without having any media tradition at all, as neither buyers of media products nor media production firms regard Rotterdam as a media location. In relation to the issue of tradition, we found that the commitment of influential firms or individuals to a city or region can do much for a cluster. For instance, the commitment of media tycoon Kirch to the city of Munich has contributed much to the development of commercial television activities in that city. In Lyons, the Boiron family, with their homeopathy conglomerate, are strongly attached to the Lyons region. In Eindhoven, partly as compensation for the move of the Philips headquarters to Amsterdam, the company decided to invest in a huge technology campus in Eindhoven. An interesting case in that respect is Leipzig, which is trying to re-establish itself as the media city that it was before the Second World War and the communist period. In Leipzig, traditional ties have survived the decades of communism: some German firms with roots in Leipzig have reopened subsidiaries to breathe new life into the ties between the firm and the city. A policy consideration of these observations is that psychological factors such as commitment and 'local attachment' should be explicitly recognised and built upon.

3.2 Critical Mass

In practice, it is difficult to determine when a cluster has 'critical mass': much depends on the definition of the cluster. In highly specialised fields critical mass can be reached with only a handful of firms and/or institutions. Nevertheless, broadly speaking the investigation confirms that large clusters in terms of the number of firms, added value and employment (such as the media cluster in Munich, the health cluster in Lyons and the telecom cluster in Helsinki), have an advantage over smaller ones due to externalities. For one thing, scale offers the advantage of increasing specialisation of firms in the cluster, thus making the cluster more complete. In Munich for instance, the media cluster includes sophisticated suppliers of digital equipment, whereas in the smaller media clusters of Rotterdam and Leipzig there is no critical mass for such specialised services. Critical mass is also needed to sustain a 'cluster superstructure', such as privately operated education facilities. An example is the Medien-Akademie in Munich, which is supported by the many TV stations. Next, critical mass entails a huge and specialised labour pool, with staff inclined to change jobs, taking best-practice and new knowledge from one firm to another and thus increasing the clusters' competitiveness, as is illustrated by the case of Eindhoven. In audiovisual activities (film, TV production) some degree of critical mass proved necessary to attract staff such as directors, actors, and camera people. The experiences of the cases point to a self-reinforcing development: a large-scale cluster entails division of labour and specialisation; the large, specialised job market generates knowledge transfer; this permits further sophistication of the 'cluster product': that, in turn, may activate more demand; next, the increase in demand stimulates firms to expand, induces cluster-specific new firm creation and attracts more firms to the cluster, so that the economies of scale increase further: see Figure 12.1 for a graphical representation of this 'virtuous circle'. Nevertheless, the circle is by no means automatic. The potential danger is that success could at the same time induce sluggishness and conservatism by (key) players in the cluster.

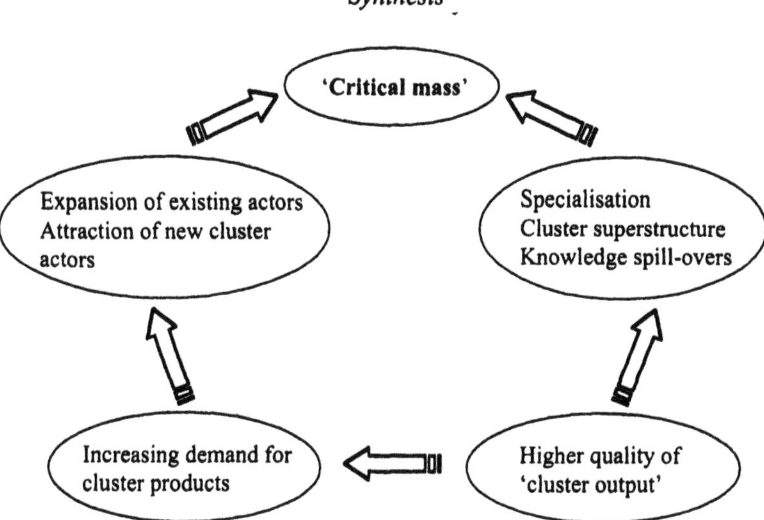

Figure 12.1 The 'virtuous circle' of cluster development

A cluster's critical mass development is also related to its market reach: the cluster's critical mass increases as the actors in the cluster develop their market beyond the local and regional market, to (inter)national markets. All of the 'mature' clusters serve the international market (for example Helsinki and Munich). Interestingly, the case of Manchester shows that actors in the cluster can develop international contacts, whereas many of the cultural enterprises still have difficulty developing the local market. For Amsterdam the tourist cluster is very much an international phenomenon; the overwhelming majority of overnight stays is attributed to foreign visitors. At the same time, the fluctuations in the international market have a profound impact on the cluster.

In sum, the clusters are in different stages of development. In order to pursue appropriate cluster stimulation policies, policy makers should be well aware of the development stage of the cluster under consideration.

3.3 Role of Large Firms

The presence of large firms in the city is a valuable asset, even if their interaction with the other cluster constituents is limited. In the case studies, we have found wide differences among big companies in the degree to which these firms are 'rooted and fledged' in the region. Some companies – such as Immuno-Baxter (world leader in the production of blood products) in Vienna and, to a lesser extent, Philips (electronics) in Eindhoven – are relatively 'inward-looking' and do not actively regard the presence of other cluster actors in the region as an advantage. Their degree of strategic networking in the region is generally small. This does not mean that these firms are unimportant: they are valuable sources of knowledge and people and a potential 'breeding ground' for spin-off firms. For instance, despite its self-sufficiency, in the mechatronics cluster in Eindhoven, the role of Philips is important as a source of high-grade knowledge (which spills over when people change jobs), as the progenitor of spin-off companies and as breeding ground for talent: many firms in the mechatronics cluster somehow have some Philips background or relation. Other large firms in our case studies proved to be more fledged in the region. Examples are Nokia in the telecom cluster of Helsinki, Novartis (pharmaceutics) and Boehringer Ingelheim in Vienna and Mérieux (pharmaceutics) in Lyons. They have linked up with universities, and provide much knowledge transfer in the cluster. In some cases, large firms even have an active policy to serve as an umbrella for spin-off firms that are not direct competitors (Novartis), from the wish to develop a set of satellite firms with complementary competencies. Not all clusters studied contain large firms: in the cultural industries in Manchester, big players are absent. The same is true of the media cluster in Rotterdam. In some cases, a cluster can become too dependent on one single firm, as seems to be the case in Helsinki, where the cluster is strongly dominated by the rapidly expanding Nokia: this firm hires more than half of Helsinki's technical university graduates; many firm in the regions are strongly dependent on commissions from Nokia. A possible downturn of such a dominant firm may have detrimental impact. The lesson is that diversification is important, both within a cluster and in a city as a whole.

3.4 Strategic Relations among Cluster Actors

In the case studies, we have laid much stress on studying relations among the cluster actors, from the idea that strategic interaction allows actors to access one another's resources (which can be markets, financial means, people, knowledge, networks) and thus contributes to the cluster's good functioning. Moreover, strategic interaction in a cluster helps to 'tie' (international) firms to the region. In the face of mergers, acquisitions and rationalisations in many sectors (notably electronics, automobiles and pharmaceutics), an international firm is much more likely to remain in the region when it is firmly embedded and fledged. An example is ASMlithography (equipment for chip production) in Eindhoven. As this strongly networked firm is very dependent on suppliers in its vicinity, its propensity to relocate is small. Another example is Boehringer

Ingelheim, a German pharmaceutical firm with a large research facility in Vienna, which has very close ties with the University of Vienna.

Strategic linkages in clusters which emerge under conditions of target convergence, recognisable shared interested, mutual trust and (cultural) proximity help to build relationships in the cluster. We found a great variety in the nature and intensity of relationships within clusters. Several types and degrees of interaction can be discerned: see Figure 12.2.

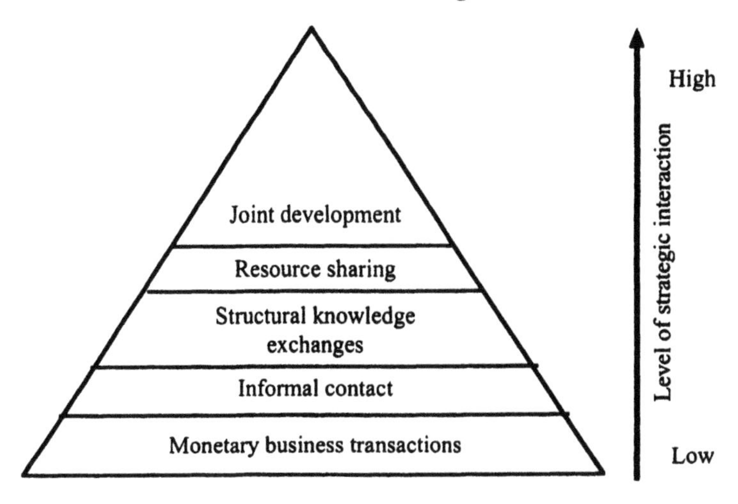

Figure 12.2 Levels of strategic interaction

On the basis of indicative evidence –interviews with business leaders, the presence of formal cooperation projects between cluster actors – we found the highest levels of strategic interaction in Munich, Helsinki and particularly Eindhoven. Amsterdam holds an intermediary position, where the actors in the field of tourism (hotels, attractions, tour operators) meet regularly to look after common interests. These actors put up a small amount of money as member of the joint marketing organisation for tourism (joint project), but joint product development takes place only occasionally. In the health cluster of Vienna, there is a considerable degree of university-industry collaboration compared to Lyons. At the lowest level, we found Leipzig and Rotterdam, where actors act relatively independently. More specifically, the embeddedness of educational institutes (in particular universities) proved to be fundamental for the cluster functioning. The 'knowledge economy' is evidently reflected in an increase of university-firm interaction. In Figure 12.3, several degrees of strategic interaction between the business community and the educational institutes are illustrated.

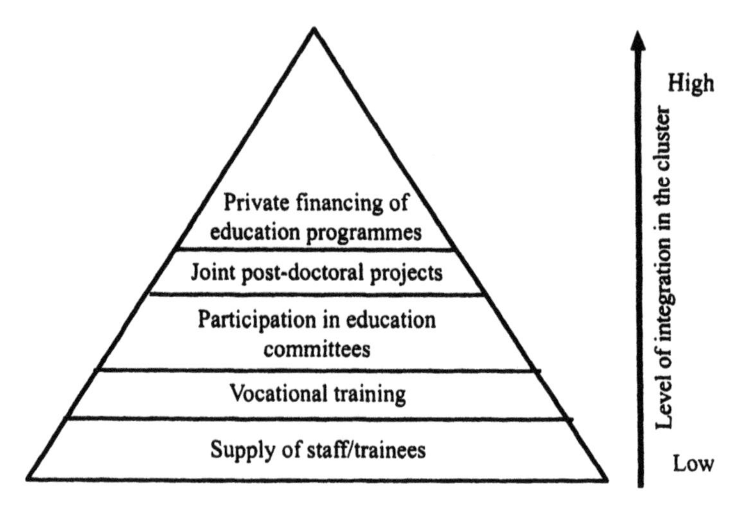

Figure 12.3 Embeddedness of education

On a basic level, the universities provide trainees and future staff for the cluster firms. In this respect we found that the match between education supply and needs of the cluster differ widely among the cases. In Amsterdam and Manchester, university education is ill-adapted to the needs of cluster firm. To a lesser degree this holds also for the health clusters of Vienna and Lyons: the firms' representatives complain about a lack of entrepreneurial skills among students and a overly one-sided emphasis on old-fashioned scientific education. The more strategic involvement of cluster firms with the university is depicted in the higher levels of the pyramid. Cluster firms can participate in education programmes (this happens, for example, in Helsinki and in Munich), use the university for vocational training or PhD projects, education for their staff, finance chairs (Philips in Eindhoven), or sponsor education programmes.

In the case studies, we have found that cooperative structures among education institutes in most of the clusters are weakly developed. In Rotterdam, three institutes offer media or media-related education on several levels, but the programmes are not compatible. A similar situation prevails in Leipzig. In Helsinki, the potentially complementary universities function in almost complete separation from each other. Our conclusion is that the prevailing 'island mentality' of many institutes means missed chances for cluster development. More cooperation – for instance in joint marketing of the city as the educational centre for a cluster, or in matching programmes on several levels – can increase the inflow of young talent into the cluster and thus strengthen its position in the future. Universities are not only important as education centres, but also as research locations. For research, just as for education, a pyramid can be drawn up that indicates the level of strategic interaction between cluster firms and research units (see Figure 12.4).

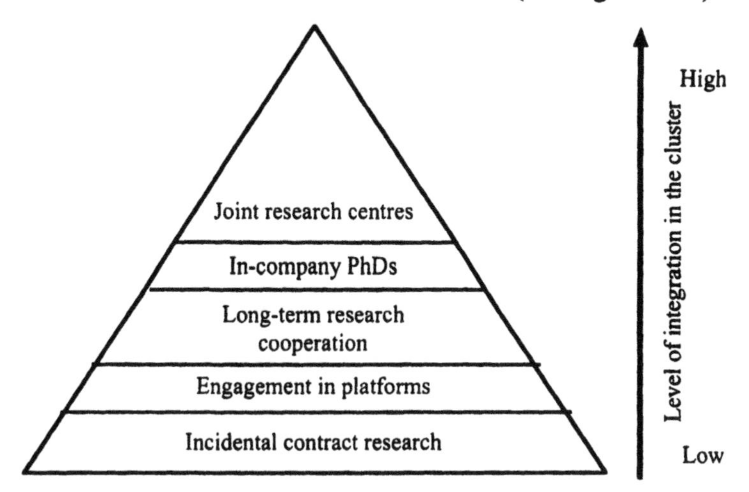

Figure 12.4 Integration of research institutes

On a fairly low interaction level, firms may incidentally outsource research, or engage in discussion/research platforms with the university. A good example is found in Eindhoven, where the university is involved in a platform on embedded systems. On a more strategic level, a university may have even more value for a cluster. For instance, the engagement of universities in longer-term contract research and licensing for cluster actors may strengthen the competitive position of firms that cooperate with the university. This holds particularly for 'research-intensive' clusters (the health cluster of Vienna and Lyons, the mechatronics cluster in Eindhoven and telecom in Helsinki). Fruitful combinations emerge readily where the more fundamental research activities of the university are a very valuable complement to the applied research of firms. The highest level of integration we found in the case studies was the joint research centre set up by the University of Vienna and Boehringer-Ingelheim Austria. Illustrative of the importance of universities for firms is the strategy of the expanding telecom multinational Nokia of locating its new research institutes (all through the world) in the close vicinity of universities. For marketing efforts of local governments to attract new firms, this implies that universities in the region should be regarded and treated as an important location factor.

An important observation is that the benefits of firm-university interaction accrue to the university as well: it generates financial resources, helps to focus research activities on matters that are relevant for business or society, and thus entails a more efficient spending of (public) money. It may also increase the quality of the research, since the demands of the market are generally high. In Lyons, where university-business interaction is at a very low level, the scientific discoveries of universities often appear useless for the business sector. A major problem hampering fruitful interaction – not only in Lyons but in virtually every cluster – proves to be the cultural difference between the business sector and universities in terms of objective orientation and time span of activities. Although university-related policies are in most cases on a national level, there might be a role for urban government to break these barriers: the potential economic spin-offs of university-industry cooperation for the region can be high.

3.5 Levels of New Firm Creation

New firms in the cluster create dynamics: they offer employment, create value added and may act as useful suppliers for existing firms in a cluster. Particularly when active in expanding markets, new firms may grow very rapidly and add even more to the cluster. New firms are started from several sources: from educational institutes, existing firms, universities (researchers who commercialise a scientific discovery), or other educational institutes. We found different levels of new firm creation in the several clusters. The clusters with the highest figures are Eindhoven, Helsinki, and Munich. Rotterdam, Amsterdam and Manchester hold an intermediate position. At the lower end, we find Leipzig, Lyons and Vienna. In our view, the level of new firm creation depends on the type of cluster, the degree and level of starter support and the general attitude towards entrepreneurship. In the medical clusters in Vienna and Lyons, to set up a new firm (for instance in biotechnology and medical technology) proved very difficult because of strict regulations, strong vested interests of existing (multinational) companies and a lack of incentives: in Vienna for instance, hospital staff have no incentive to develop new products, as any benefits of patents accrue to the city, not to the inventor. In other fields – such as software development, new media and communication – it is much easier to start a new business, because of less regulation, fewer requirements in terms of scale, technology and capital and a less mature market. In Lyons, more than in other clusters, we found that the huge cultural and mental gap between universities and business world seriously hampers the development of spin-off companies from the university. Another conclusion is that support for starting firms should not remain restricted to financial support and space provision, but become more integral and more targeted. A good example of an integral approach is the starters' facility in Munich, which offers not only office space and all kinds of support but also offers starters access to networks of established firms in the region. The concept of 'twinning' new firms to existing ones is also developed in Eindhoven, where large firms contribute to a starters' facility, not only financially but also by sharing their knowledge and networks. In other clusters as well (for instance in Vienna), large firms have recognised the benefit of the proximity of young, dynamic complementary firms and are willing to invest in it with several resources. From this, we may conclude that starters' policy should not be a matter of public agencies only: precisely the knowledge, experience and networks of existing firms can make a starters' policy successful and should be used to the full.

4 Organising Capacity

The final element that we presented in the analytical framework as one of the factors that contribute to the development of growth clusters is the degree of organising capacity regarding the cluster. Research has identified several factors that contribute to organising capacity in cities. In this investigation we have focused on vision and strategy, strategic public-private interaction, leadership and political and societal support. In general, we found that a high degree of organising capacity can be of great help to identify chances, to design and effectively implement strategies and to prevent wrong investment decisions.

4.1 Presence of an Integral Vision on Cluster Development

The experiences of the cities illustrate that some successful clusters are supported by an integral vision on the development of the cluster in the context of the local and regional economy. The city of Amsterdam has a clear vision of and strategy for the tourist cluster, broadly supported by key actors in the cluster itself. Eindhoven has made the promotion of networking and partnerships in the region a leading principle in the region's economic policy. This is particularly important for the mechatronics cluster in which the combination of different technological disciplines is essential. In the case of media in Munich, it was the Free State of Bavaria that developed a policy favouring the media cluster. Ten years ago the city of Munich was not very supportive of the cluster, but that attitude is changing, with positive initiatives such as the Munich Technology Centre, as a sign of the new strategy in the city. The approach to the cultural industries in Manchester is also changing. Culture and cultural enterprise has been given a place in the region-wide regeneration strategy, with Manchester City Council now working on a policy scheme for tailor-made support to cultural business in the fields of design, media, multimedia and popular music. Lyons has developed an integrated vision on the health cluster with five concentration poles, however, there is no clarity on the development direction of these focal poles. In the other cities a clear, fully balanced vision of and strategy for the development of the cluster as a whole is yet to be developed or is in progress (Vienna, Rotterdam, Leipzig and Helsinki).

4.2 The Role of Strategic Public-private Cooperation

In the previous section the relationships among actors within the cluster have been centrally placed. The investigation has also included strategic interaction between the public and private sectors concerning the development of cluster strategies. The case of Rotterdam serves as an example of where the lack of strategic interaction between the city departments and the business community has resulted in ad hoc policies, where projects such as the 'Schiecentrale' and the Van Nelle building are developed independently from one another. In Manchester the strategic contacts between the city and the cultural industries could be improved as well. The city's cultural industries are an economic factor as well as source of creativity that the city government could use in the marketing of Manchester. Lyons is an example of good cooperation between the Communauté Urbaine and the Chamber of Commerce in stimulating development in the health cluster. The structural consultation between key figures in the Helsinki club is a positive development in Finland's capital and might lead to efforts to overcome the lack of a metropolitan vision also with regard to the telecommunications cluster. The strategic interaction in the mechatronics cluster in Eindhoven has been strongly stimulated by the Stimulus Programme, leading to public and private investment in the Twinning Centre, whose aim is to accommodate young entrepreneurs and twin them with the expertise of senior business people. It can be concluded that public-private cooperation is a prerequisite to developing effective and efficient cluster policies. 'Interactive policy making' is needed in the marketing of the cluster, in attracting new firms, in helping start-ups and in all other aspects of cluster policies, to make optimum use of the knowledge and resources of the existing actors in the cluster. This also implies that civil servants involved in cluster policies need to be well educated and have sufficient empathy with the cluster.

4.3 Political and Societal Support

Political and societal support is a necessary condition for cluster development. Clusters with growth potential are helped by well-developed political and societal support and lack of support is a threat to growth possibilities for the cluster. One of the clearest examples is the case of tourism in Amsterdam, where tourism causes inconvenience to inhabitants, in particular those in the city centre. There is still enough political and societal support, but the challenge for policy makers is to sustain support as the cluster continues to grow. The spread of visitors across the region is also vital to maintain support. The promotion of the media sector is supported wholeheartedly in the political circles of Leipzig and can count on support from the population as well, since unemployment is still a major problem for the city in transition. In Vienna, the negative attitude of the general public towards gene manipulation hampers (public) investment in starter facilities in biotechnology, one of the most dynamic parts of the health cluster.

4.4 Leadership

Leadership in cluster development comes in different disguises. In some cases, the major firm(s) in the cluster, such as the pharmaceutical firms in the health clusters in Lyons or in Vienna, and Nokia (basically because of its size not out of strategic considerations), take on the leading role. In the cases of Manchester and Rotterdam such leadership by major firms is absent in clusters in which the firms are predominantly small and medium-sized. There is certainly a case for public leadership in cluster development, to establish missing links in the clusters, to promote new technology or to create incentives for cooperation. In the Munich case, the Free State of Bavaria has been the leading public actor in the field of the media: it took the initiative and created conditions to attract commercial TV stations and promotes the use of new media. In other cases, stronger public leadership could set development in motion. In Manchester, urban management could stimulate linkages between cultural enterprise and educational institutes. In the case of Lyons it is the strategic interaction between the universities and the business community that could be improved. In Rotterdam, the city could stimulate cooperation in the scattered audiovisual sector to become a serious interlocutor for potential principals.

5 Concluding Remarks

This international comparative research has shown that the cluster perspective in the studying of growth processes in cities has added value. Increasingly, economic activities cross the boundaries of traditional economic sectors, as networks are becoming the leading organisational principle. These networks are the vehicles for new combinations, innovation and growth. Urban competition requires cities to make optimum use of their resources. As resources are dispersed among many actors, strategic interaction among (semi)public and private actors has become vital. In this light, for urban economic policy, awareness of the dynamics of cluster development is essential. A cluster approach is a useful policy tool: it sheds new light on relevant interlinks among key economic players and may thus bring about renewal and innovation. Furthermore, cluster-oriented policies can be a means to tie firms to the region. An effective cluster policy calls for a broadly supported vision on the development potential of the cluster in the context of the regional economy. The investigation has illustrated that cities in the several European countries face similar challenges as they are all seeking to capitalise on new growth opportunities. Indeed, urban regions are the economic engines of Europe, as major concentrations of all kind of economic activity and as breeding grounds for innovation and renewal. Although there is no explicit urban policy at the European level, the awareness of the European Commission of the great economic value and potential of cities is increasing. This study indicated that in economic growth and innovation processes, the local and the regional levels are gaining importance. Despite growing degrees of internationalisation, and even globalisation, we found that urban regions are important spatial and functional units: many firms derive their strength from their 'embeddedness' in regional networks, while at the same time they are active internationally. A first step of a more active engagement at the European level could be the structured facilitation of information and best practice exchange and the sharing of experiences, regarding growth processes and dynamics in European urban regions. This is all the more important as growth and innovation are not aims in themselves, but necessary to counter unemployment and to remain competitive in the longer run.

For Product Safety Concerns and Information please contact our EU representative GPSR@taylorandfrancis.com
Taylor & Francis Verlag GmbH, Kaufingerstraße 24, 80331 München, Germany

www.ingramcontent.com/pod-product-compliance
Ingram Content Group UK Ltd.
Pitfield, Milton Keynes, MK11 3LW, UK
UKHW021442080625
459435UK00011B/345